Application Service Provider and Software as a Service Agreements Line by Line

A Detailed Look at ASP and SaaS Agreements and How to Change Them to Meet Your Needs

Kelly L. Frey Sr.
Thomas J. Hall

Mat #40705185

ISBN 978-1-59622-853-5

Library of Congress Control Number: 2007941891

For corrections, updates, comments or any other inquiries please email AspatoreEditorial@thomson.com.

First Printing, 2007

10 9 8 7 6 5 4 3 2 1

Aspatore Books is the largest and most exclusive publisher of C-Level executives (CEO, CFO, CTO, CMO, partner) from the world's most respected companies and law firms. Aspatore annually publishes a select group of C-Level executives from the Global 1,000, top 250 law firms (partners and chairs), and other leading companies of all sizes. C-Level Business Intelligence™, as conceptualized and developed by Aspatore Books, provides professionals of all levels with proven business intelligence from industry insiders—direct and unfiltered insight from those who know it best—as opposed to third-party accounts offered by unknown authors and analysts. Aspatore Books is committed to publishing an innovative line of business and legal books, those which lay forth principles and offer insights that, when employed, can have a direct financial impact on the reader's business objectives, whatever they may be. In essence, Aspatore publishes critical tools—need-to-read as opposed to nice-to-read books—for all business professionals.

CONTENTS

1

Introduction

ASP and SaaS—NIAs (Non-Instructive Acronyms)

The world of information technology is filed with acronyms, which, while distinctive, are not always particularly instructive. Such is the case with application service providers (ASPs) and software as a service (SaaS). These terms were initially coined to describe technology. They have now come to represent evolving technology business models.

ASPs developed to support applications for those customers who could not economically justify maintaining such applications (and the infrastructure to support such applications) internally. Some of the earliest ASPs were the computer service bureaus of the "mainframe" era that essentially "rented" computing power to companies who could not afford to purchase or maintain a "computer" or which automated processes that had previously been performed manually. These early business models involved access to:

1. Large volumes of digital data (such as the early databases accessed through a dedicated dial-up modem)
2. Specialized software applications that were only used intermittently (such as complex actuarial calculations that were run on a monthly schedule)
3. Bulk transaction processing that automated previously manual processes (such as batch processing of invoices and billing)

As computers, computer applications, and computer expertise became more available, businesses tended to shift automation of business operations in-house, leaving the ASP field to highly specialized applications (such as database research and low volume/high value business operations).

As computing hardware and software became more accessible to smaller business, this trend toward ASP specialization continued at an accelerated rate.

Then a confluence of unrelated events created new opportunities for the ASP model:

- The personal computer became ubiquitous.
- The redundant digital communications technology developed by the Department of Defense became commercially available as the Internet.
- The telecommunications industry created a global high-bandwidth wired and wireless network in anticipation of capitalizing on a consumer-level communications revolution.

These events provided the ASP industry with a huge "installed base" of access points, an exponentially increasing base of consumers comfortable with personal computers (and their applications), and cheap connectivity at commercially reasonable transmission speeds. ASP providers also capitalized on offshore manufacturing and Moore's Law to offer more cost-effective solutions as consumer demand increased beyond what most internal capital budgets could support on an in-house basis.

As a result, businesses began to reevaluate whether it was efficient for them to support all their business applications with a vast in-house infrastructure. Server farms began to proliferate. ASPs began to port common business applications to Web browser technology. Customers began to critically examine the economics of carrying capitalized computer equipment and application development costs on their balance sheets rather than outsourcing them as expense items. These developments created the current environment for the traditional ASP model.

Coincident with these developments, a more highly leveraged ASP business model developed for niche applications (such as credit card payment systems, health care practice applications, customer relationship management support, etc.). In this more highly leveraged model, providers offer a standard product/service designed specifically to support multiple customers via Web-native architecture. Such a configuration (a "multi-tenant architecture") allows providers to serve a huge number of clients simultaneously. This model is sometimes referred to as "software as a service" (SaaS), implying a provider's ability to scale capacity to meet customer demands for "service" without significant investment in additional infrastructure on the part of the provider.

SaaS providers are typically reluctant to negotiate terms/conditions or to customize applications, as their pricing model is based upon providing standardized services to large numbers of clients. Some of the larger SaaS providers merely provide Web-based "click through" licenses (e.g., Salesforce's customer relationship management applications).

The Model Contract

The model contract we will discuss (the "model") is designed for use by a large service provider, with extensive support capabilities, when providing a business-critical service to a sophisticated customer (who may require specific customizations). As such, the model is usually subject to significant negotiation (which will be covered in the line-by-line analysis, along with alternative clauses favorable either to the provider or the customer). Alternative versions of the contract suitable for general purpose or commodity SaaS environments can be reduced to "click-through" agreements, such that the potential customer can review the terms and conditions of service and then affirmatively consent to such terms and conditions with a mouse click prior to being provided access to the service.[1]

The model provides that the provider will have significant service level obligations, such that customer is assured that its "business-critical" services continue to operate reliably. These obligations are set forth in service level agreements (SLAs), which tend to be characterized in terms of:

1. Response time after an alert from the customer of a service failure
2. Repair time to have the service operational after a failure

---

[1] For a more complete analysis of click-through agreements in this context, see *Selling Products and Services and Licensing Software Online: An Interactive Guide with Legal Forms and Commentary to Privacy, Security, and Consumer Law Issues*, ABA Business Law (2007).

The more demanding the SLAs are, the greater the risk to the provider and the greater the expense of maintaining appropriate equipment and personnel to remedy those problems that inevitably occur. During negotiations, a customer will usually request the "quickest" responses possible. In anticipation of such requests, most providers have a rate card setting different prices for different service levels. Customers may then pick the combination of price and service level that is appropriate for their specific business environment. For ASP services that involve outsourcing of infrastructure, the customer may also require specific physical or technical parameters (e.g., connectivity at a certain transmission speed, servers in a specific configuration, physical separation of a customer's processing, etc.).

The model also anticipates that the provider is able and willing to maintain multiple, customized versions of the services for its customers. Not all providers offer this flexibility, for such flexibility reduces the scalability of their business model. Alternatively, the provider may create a limited number of standard offerings available to all customers, perhaps at different prices for different types or quantities of service. In this way, providers create a flexible offering intended to appeal to a larger customer base (i.e., more powerful service offerings for large companies at increased prices) while maintaining a manageable number of operating environments. This difference is one of the key distinctions between ASPs and SaaS; SaaS offerings typically do not provide for customization but operate standardized platforms that are available under similar terms and conditions to all customers, allowing for economies of scale to the provider in hosting and pricing the services.

The Value Proposition

An attorney can add value to an ASP/SaaS transaction at two distinct points in the negotiation process.

The initial value an attorney provides to an ASP/SaaS provider is in drafting a contract that appropriately *implements the Provider's business model.* From a provider's perspective, the most economic offering may be a single application available to all customers on a "best efforts" basis. If that is the provider's business model, then contract drafting becomes a matter of "giving a little" and "reserving a lot." In most cases, such providers will prefer Web-based, "click-wrap" contracting to accommodate larger economies of scale (and reduce the transaction cost of customer acquisition that is involved in extended contract negotiations). These "best efforts" service arrangements are usually targeted at small business, local business, specific vertical markets, services for which the provider may be the sole source, or for providers advocating a "low-cost" solution. If the provider's business model is based upon service to larger, national customers or for applications that may be "mission-critical" or "business-critical," then "standard terms and conditions" have to accommodate the more exacting requirements these types of customers maintain as a matter of corporate policy (for example, larger customers outsourcing business-critical applications will require assurances that the ASP has sufficient redundancy to avoid a catastrophic failure that would halt the customer's business operations).

The on-going value an attorney provides to an ASP/SaaS provider is in *maintaining the Provider's business model* over time and over numerous contract negotiations. Even if a provider has selected a "single offering" business model, over time customers may gradually press for different terms and conditions to accommodate the specifics of each of their unique business environments. Such "accommodations" may be significant (such as supporting another operating environment, allowing more users than anticipated, customizing an interface for the customer, etc.). While these accommodations may seem appropriate in the specific circumstances (and the customer revenue may be sufficient to warrant consideration of such accommodations), the economics of the original "single offering" business model cannot support them, and the provider will lose the economies of scale inherent to that preferred model. Custom interfaces mean custom patches and custom support. Multiple support requirements may mean multiple virtual or physical platforms and cross-training of support and maintenance staff. Increasing the number of licensed users beyond the scale of the single offering platform can lead to deterioration of the service for all customers. These types of accommodations lead to dissatisfied customers and potential plaintiffs.

From a customer perspective, understanding which elements in the contract are "business model"-related allows the attorney to focus negotiations on those areas where compromise and creative solutions are possible (rather than the usually unproductive discussions of why the provider's business model should be changed at the customer's request). Also, from a customer perspective, there may be some dynamic tension that eventually surfaces during contract negotiation between (i) in-house technology business clients seeking to obtain cost savings by outsourcing applications as part of corporate strategy and (ii) in-house compliance and policy staff tasked to ensure business continuity and implementation of enterprise-wide standards of service. The negotiation process will attempt to rationalize these sometimes conflicting goals.

Whether representing the provider or the customer, most of the effort in documenting ASP and SaaS arrangements relates to maintaining the absolute restrictions of the provider (i.e., the elements critical for the provider to maintain its business model) while trying to accommodating the actual (as opposed to perceived) business requirements of the customer. Effective negotiations involve educating both sides to the absolute requirements, while trying to appropriately allocate the risks involved with the service itself. For example, the provider's business model may have an absolute restriction such that the provider cannot dedicate a physical server to the customer (so the customer's digital information is physically separated from all other customers). If that is an absolute threshold attendant to the provider's business model, then, rather than dealing with "why the customer's operating environment and policies require physically separated processing," the more constructive discussion becomes what non-physical safeguards can be implemented by the provider that provide relatively the same risk profile to both the provider and the customer (i.e., logical separation of data, controlled access to a partitioned server, audits regarding data security, encryption capabilities, etc.). Another simple example involves SLA response times. Unless a provider's business model accommodates the cost of providing support twenty-four hours a day and seven days a week, discussions with a customer that needs such support may not be fruitful.

Working with the Client

When providing legal counsel to a provider, the first discussion is about the business model. What is the preferred business model? What flexibility does the provider have with respect to the model? What type of unique agreements has the provider entertained in the past? How have these types of unique relationships worked?

After business model elements, the discussion turns to the negotiation process and the procedure for escalating questions within the client's management structure. An advocate needs to know which representatives of the client have authority to grant concessions with respect to which elements of the agreement. It may also be helpful for the advocate to document the escalation process in playbooks or manuals for use by the client's staff. In that way, the knowledge the advocate develops will become "corporate knowledge" of the client, thereby reducing legal expenses and increasing the opportunities for internal staff to contribute at a higher level. An advocate should also seek to determine which points cannot be compromised (e.g., unlimited liability for patent infringement may be a non-negotiable term, or naming a customer as an "additional insured" on the provider's insurance coverage may not be administratively possible).

When representing a customer, the attorney should strive to understand how the service fits into the client's business operations. If the service is business-critical, the advocate will want to take appropriate steps to ensure the client's business continuity. Such steps may include an outright license from the provider to the customer such that the customer may install and use the software at the customer's site if the provider is unable or unwilling to perform the services as contracted, or an escrow of the software with a third party (that is regularly updated) to ensure that, if the provider is compromised, the customer will have practical access to the application. Building legal remedies into the contract for the customer is a basic requirement. But if the service or application is critical, no amount of damages, collected long after the initial service failure, may be enough to overcome the actual losses sustained by a client who cannot duplicate the service in-house. Thus a customer's advocate should seek to build in as many practical protections as possible, attempting to ensure that customer will continue to have reliable access to the service or application, no matter what happens to provider, or to the relationship between customer and provider.

The advocate for a customer will also want to make sure the services and service levels are defined specifically *and* objectively, such that they are both practically and legally capable of being enforced. That means avoiding ambiguous terms such as "commercially reasonable" in favor of explicit contract requirements (e.g., "using commercially reasonable efforts to respond within two hours" becomes "restore service within two hours"). From a customer's perspective, the provider's "efforts" mean nothing. What counts for a customer is the provider's actual performance to contract specifications.

2

Line-by-Line Analysis
of a Master Subscription Agreement

Opening Information

In most ASP or SaaS agreements, the first paragraph recites a variety of factual details, such as:

- The name of the agreement
- The names of the parties
- The date of the agreement

Each of these details may become a defined term. For example, "Master Subscription Agreement," could be defined as the "Agreement." Similarly, the date of the agreement may be the "date of execution" or may refer to the date upon which the agreement formally takes effect (i.e., the "effective date"). This date then becomes a defined term for the purpose of both:

1. Uniquely identifying the agreement in question;
2. Establishing the time periods during which the agreement will be in force.

The date on which the agreement takes effect is not trivial and may be negotiated. This critical date is typically reduced to a defined term in the agreement (i.e., the "Effective Date" or "Contract Date").

The opening information should also include the full legal names of the entities to be bound. These names can be abbreviated for convenience into defined terms (such as "Provider," or "XYZ" as an abbreviated form of "the XYZ Corporation"). The drafter needs to carefully distinguish between the "party to be bound" and the entities that have the benefit of the agreement, particularly if any of the parties are large companies or members of large corporate organizations. For example, a corporate parent may be the "party to be bound" (that is, responsible for legal performance of the agreement), but the opening information may include references such as "the XYZ Corporation, for the benefit of its wholly owned subsidiaries." Such a provision allows the parent to contract for services (and contain the liability for performance/payment for services at the parent level), but allows the subsidiaries to obtain the full benefit of the negotiated agreement. From a provider's perspective, while it may be preferable to have a parent contract for all of its corporate affiliates, the provider must ensure that:

1. The terms of the agreement are enforceable against affiliates that will ultimately benefit from the agreement; and
2. The affiliates that have the benefit of the agreement can be objectively determined (for administrative and enforcement purposes).

Enforceability is especially important for the provider with respect to nondisclosure of confidential information and trade secrets. Note that the customer may specify that responsibility for performance resides with the parent in the provider's organization, to ensure that the entity actually performing the services under the agreement has sufficient resources and that there is a solvent defendant (in the event of a dispute).

While providing mailing addresses in the opening information may be helpful, there should typically be a "Notice" provision in the miscellaneous section that defines where notices relating to the contract should be directed. Jurisdiction/venue/choice of laws provisions should be explicitly set out in the miscellaneous section (so casual references to locations of multi-jurisdictional parties in the opening information do not create an environment that encourages "forum shopping").

This Master Subscription Agreement ("Agreement") is made and entered into as of [insert effective date] (the "Effective Date"), by and between [insert Name of Provider] ("Provider"), a [insert

state of incorporation] Corporation, with offices at [insert location of offices], and [insert Name of Licensee] ("Licensee") a [insert state of incorporation] Corporation, with offices at [insert location of offices] ("Customer").

DEFINITIONS

The definitions section is one of the most important, yet least utilized sections of most agreements. The definitions section is more than just a "glossary of abbreviated terms." When used with care, this section can "define in" or "define out" critical service components and legal requirements. For example, the definitions section typically addresses ownership of intellectual property or types of information that will require special handling (e.g., what information is considered to be confidential).

If expansive or non-standard services are to be provided, best practice is to define these services by reference to exhibits or schedules that set forth the technical specifications in detail. This technique allows the "business client" to set forth fully the exact technology/application to be provided, as well as the specific service levels that are expected in support of the technology/application (e.g., "the Gold Service Level, as set out in Provider's April 26, 20__ technical specification manual, which technical specification is incorporated herein by reference").

On a practical level, the provider will want to define its obligations as narrowly as possible, and will attempt to add as many disclaimers or qualifications as possible (e.g. "provider will use *commercially reasonable* efforts to ensure service at the *prevailing industry standards*"). The customer, on the other hand, will typically want to use the definitions section to expand the affirmative obligations of the provider and to create objective standards for performance (e.g., "provider will provide the service on a 24/7 basis, subject only to the scheduled maintenance and *force majeure* provisions of the Agreement set out in Sections [__] and [__], respectively").

If not explicitly articulated elsewhere in the agreement, the drafter should provide that the terms defined in the definitions section will take precedence over any other provision of the agreement or a subsequent schedule in the event of a conflict or internal inconsistency in the agreement.

The model is denominated a "master" agreement; it is intended to govern multiple services/engagements provided over time to the customer. It is designed so such services can be defined both within the body of the agreement and its schedules/exhibits, and in subsequent statements of work (SOWs). This design provides flexibility for ongoing relationships and reduces the transaction costs of adding services (all subject to the basic legal terms and conditions) over the term of the agreement. With respect to SOWs, master agreements often specify that the terms of the master will control over any conflicting terms in any statement of work. If greater flexibility is needed, master agreements sometimes direct that the terms of the master agreement can be modified within a statement of work, provided the statement of work specifically identifies the section of the master agreement being changed (e.g., "Section X of the Master Agreement shall not apply to this Statement of Work.") This structure provides flexibility over the term of the relationship, while minimizing inadvertent "scope creep" as multiple SOWs with wording in conflict with the master agreement take the parties further away from the original terms of the agreement.

Best practice when using defined terms is to vary their physical appearance (at least the first time they are referenced in the agreement). Conventions vary from putting the defined terms in parenthesis and quotes (as most attorneys prefer) to adding italic and bold font style to distinguish defined terms from text.

Definitions

1.1 "Consulting Services" will have the meaning set forth in section 3.1.

If the customer is merely subscribing to an "off-the-shelf" service, consulting services may be a minor part of the transaction. On the other hand, if consulting will be a significant portion of the contract, those services should be defined with care and with reference to "dates" and "deliverables" such as:

- What services will be provided?
- When?

- Where?
- What will be the performance standards?
- Who will provide the services?
- At what price?

1.2   "Confidential Information" will have the meaning set forth in sections 5.1 and 5.2.

"Confidential information" should be carefully defined to protect information that is not generally known and which is either protected by law (such as certain types of health records) or commercially valuable. Avoid broad definitions, such as "Confidential Information means all information disclosed by one party to the other." Also, in anticipation of any term limit on nondisclosure of confidential information, remember that trade secrets may lose protection unless they are protected from disclosure in perpetuity.

1.3   "Customer Data" means all information provided by Customer to Provider through the Service for use in conjunction with the Services and the Software, including processing, storage, and transmission as part of the Services.

1.4   "Customer Information" means all information created or otherwise owned by Customer or licensed by Customer from third parties, including Customer Data and information created by Customer by using the Services, that is used in conjunction with the Services and the Software.

1.5   "Documentation" means all configurations and specifications published by Provider from time to time relating to the Software or the Services.

This definition of "Documentation" focuses on the technical specifications for the software. The materials specified would be of most value to personnel providing maintenance and support. The end user, however, might not find such materials helpful. A careful drafter will therefore take the time to determine what documentation exists, and what is needed, and then specify the customer's rights to it. For example, assume "documentation" includes user manuals. How many copies will the user receive? May the customer make additional copies for its own internal use? Or, if the customer creates specialized materials, documenting how the customer uses the provider's services or systems, who owns those materials, the customer or provider?

1.6   "Equipment" will have the meaning set forth in section 4.2.

If the focus of the contract is on the type and quantity of services to be provided, it may not be necessary to spend much time defining equipment. A provider attempting to leverage a standardized infrastructure may reasonably be expected to resist requests for specialized equipment. Thus a customer may be better served by focusing on service levels and insisting upon special equipment only if absolutely necessary. Equipment definitions are typically important where specialized equipment is necessary for the specific customer usage or where the definition of equipment will be helpful to the customer should it have to replicate the provider's hardware environment.

1.7   "Locations" means the physical location or locations set forth in Exhibit A from which Customer is licensed to access the Service.

Will the customer be permitted to access the services from one location, half a dozen, or from anywhere an employee can access the Internet? Customer and provider should explore these questions thoroughly, and document the answer in their contract.

1.8   "Maximum Users" will have the meaning set forth in section 12.1.

"User" can have specific legal and/or economic significance. For example, will "ten maximum users" mean ten named employees (e.g., John Doe, Jane Roe, Sally Smith, etc.) or any ten employees, so long as there are never more than ten individuals using the services at one time? Beyond that, will the customer be restricted, now or in the future, by limiting the customer to a maximum number of users? If the service is priced by the "number of users," does that term imply unique individuals, the number of passwords, simultaneous users, etc.?

1.9 "Provider Information" means all information, including the Software, created or otherwise owned by Provider or licensed by Provider from third parties, related to the Services and any materials prepared by Provider pursuant to a Statement of Work under the Agreement.

The provider will, of course, want to protect all of its materials. On the other hand, the customer might question including this term unless such information will be provided to the customer. The customer may also question including "materials prepared by Provider pursuant to a Statement of Work" in the definition of provider information. The customer may wish to own the rights to such materials, or the materials may contain information that is confidential to the customer, or a trade secret, or be protected by privacy laws. The drafter may want to split this term into its various components, such as "provider software," and "provider confidential information," while addressing work product as a separate term.

1.10 "Services" means Provider's electronic data processing, storage, and transmission services ordered by Customer, which are enumerated in Exhibit A.

In the model, the "services" to be performed are defined by reference to a section in the agreement or a schedule to the agreement. Since the affirmative obligations of the provider are typically related to the services, this definition articulates the true legal obligation of the provider to the customer. To ensure that a court is capable of enforcing the agreement, the services must be defined objectively.

1.11 "Software" means the software used by Provider to provide the Services.

As with equipment, a customer seeking a certain level of services might be better served by defining the type and quantity of services rather than attempting to define the software the provider will use. The model, however, presumes that the customer seeks a certain application, rather than generic functionality. Additionally, should the customer require that the application be escrowed with a third party, the exact definition of software will define the materials provider must deposit with the escrow agent.

1.12 "Statement of Work" will have the meaning set forth in section 3.1.

SOWs are a standard tool for adding transactions under the umbrella of a master agreement. When laying out the format of an SOW, care should be taken to ensure that:

- The SOW becomes effective only upon execution by appropriately authorized representatives for both sides;
- The terms of the master agreement control over the terms of any SOW, unless an explicit change is agreed upon in the SOW and the master allows for such change;
- The SOW sets out services, pricing, and other responsibilities with the same care as the master.

Access to Services/License to Application

Although ASP/SaaS services are generally delivered via the Internet, the services themselves rely on an array of software and hardware. This fact creates a variety of challenges for those who negotiate and draft ASP/SaaS agreements. For example, is the customer granted the right to run a specified/defined number of transactions through the provider's system in a month, or is the customer granted a more general license to use the software embodied in the system without any volume limitations? SaaS agreements generally focus on the former approach and refuse to grant customers any rights in the provider's technology (this is especially true of providers who offer mass-marketed services using click-through agreements). In contrast, the model focuses more on the typical ASP transaction, in which a provider may offer application services, some consulting, some customization, and some license rights to the underlying technology (or at least to the application software).

With regard to services and support, the model is neutral to provider-friendly. Generally, the provider is expected to make "reasonable commercial efforts" rather than being required to meet a specifically defined service level. Note that there is no accepted technical standard or definition for "commercially reasonable effort." If a dispute develops, the definition may be determined by a judge, but only after much time and expense. Consequently, astute customers generally do not accept the "commercially reasonable" standard. The more important the service is to the customer's day-to-day operations, the more important it is to define the required level of performance in detail and in terms that can be objectively measured.

The provider has the option of providing additional services and upgrades, and may require the customer to use the upgraded versions. Astute customers typically resist such provisions. Upgrades by the provider may require new fees, reliance on new, untested hardware or software, increased training costs, or any combination of the three.

Drafters must bear in mind two competing themes: providers seek to maximize current and future income, while customers seek to control current and future expense. At the same time, each side is attempting to minimize its potential legal and business risks.

In addition, the more important an application or service is to the customer, the more the customer will resist giving the provider the right to make unilateral changes. Imagine if one morning the customer discovered the payroll system no longer worked, or no longer operated in the manner to which the customer was accustomed. Thus a customer may insist on language such as:

> *"Provider shall retain Version 1.0 of the Software for the term of the Agreement and shall not utilize any other software, or any other version of the Software, to provide the Services to Customer without Customer's prior written consent."*

Or, to give the provider somewhat more flexibility:

> *"Provider may from time to time implement upgrades to the hardware or software used by provider to provide the Services, provided that the functionality of the Services shall not at any time fall below the levels specified in the Agreement or any applicable Statement of Work."*

Should training be required for the customer to upgrade during the term, costs of such training may be a negotiated term (this alternative is not included within the model). Note that a written request for additional services will typically be required to be in a specific form/format (and the parties may provide for an additional exhibit or schedule to the agreement that sets out that form/format).

But the model is not completely pro-provider. For example, in the model the customer receives an explicit license to the underlying software (although that license falls far short of the type of license most large corporate clients would require).

Most license provisions in ASP agreements favor the provider. They "give a little and take back a lot." In the model, the explicit license is limited, non-transferable, non-exclusive, and solely for the term of the agreement. It is further constrained by the requirement that the customer use the services only from defined locations and solely for the customer's "normal business." Care should be taken with such terms as "normal business," as everyone presumes they know what the term means—until a dispute arises. For example, ACME subscribes to the Mega Payroll system for its "normal business." Does that mean ACME may use the system to run only payroll for ACME employees? What if ACME is a service bureau and makes its money by selling access to computer programs and systems? In the long term, both provider and customer are better served when they take the time to carefully define what are, and what are not, permitted uses.

A more customer-friendly license grant would expand the license to an enterprise-wide scope, while the number of end users could be varied (increased) over the term of the agreement merely by paying appropriate fees. The license would also run to the customer and possibly to all of the customer's affiliates, agents, business partners, and so on (i.e., anyone acting on behalf of the customer who may need the services). Other provisions of the model would need to be conformed to such an expansive license (e.g., the restriction that the software be used only at one specified location could conflict with the needs of agents of the customer, who might require remote access). Location restrictions are not customer-friendly and impose administrative burdens and costs on both parties. They should be avoided unless critical to the provider's business model.

As the provider's application becomes more business-critical and the customer becomes larger (or subject to regulations or policies regarding business continuity), additional customer-side provisions may need to be negotiated. These provisions might include:

- **Divested entities:** Assuring that the license grant and use of the services are extended to all affiliates of the customer, even after they are no longer affiliated with the customer (at least for a time sufficient for the divested entity to negotiate its own license with the provider without an interruption of business operations).

- **Source code escrow:** The requirement that the provider place a copy of the application (and documentation sufficient for the customer to install and maintain the application) with a third party escrow agent. Such escrow agreements typically:

  o Detail when the source code and documentation may be released to customer;
  o Require that the provider update the escrow deposit as the application is modified or updated;
  o Permit customer to have the escrow deposit verified by an independent third party.

The customer will also need a license to host, maintain, and modify the application after obtaining the source code (although the provider will typically insist that such license be limited, so the customer may not remarket or resell the application.) Providers typically resist escrow agreements, as they do not wish to risk losing control of a vitally important asset—their software. Thus providers typically insist that the source code be released only in extreme conditions. Conversely, customers generally do not want the burden of operating the software or the application (since that is a burden the customer was trying to avoid when it contracted with provider.) For the customer, therefore, access to the source code and documentation is a remedy of last resort.

- **Transition services:** Such that in the event of any termination, the customer's business operations are maintained until a substitute for the services can be obtained.

- **Patent license:** A license to any underlying patentable invention incorporated within the services or software (or at least a non-enforcement provision against the customer during the term of the agreement).

Consistent with "give a little, take back a lot," most definitions of services will also contain negative (as well as positive) rights for the customer. In the model, these negative obligations relate to maintenance of the competitive advantage of the provider and assurances that the customer will not become a competitor to the provider. Additional affirmative obligations of the customer might include:

- Compliance with relevant laws;
- Policing customer's use of the services to ensure that no third party has unlicensed access to them via the customer's internal systems

While such restrictions may seem trivial, they can create significant financial responsibility for the customer. Specifically, they require a present commitment to avoid certain business operations in the future (usually a decision made at the level of a customer's board of directors, not by the technology procurement staff) and create at least an evidentiary burden for the customer, should it, in the future, independently develop a functionality similar to the services (i.e., proof of independent creation versus copying or use of the provider's intellectual property). Such provisions may also create goodwill issues if a customer has to enter into actions initiated by provider against unauthorized access/use by third parties claiming rights to use through the customer (i.e., a customer grants usage rights to a business partner, who makes more expansive use of the right than granted—the customer is now forced into the position of "plaintiff" against its business partner under the terms of the agreement with the provider).

In addition to defining the services, the model specifies what support will be available in the event of an interruption or failure of the services. The precise type and level of support is defined in schedules or exhibits attached to the model. Each schedule should detail what services will be provided, when, and how quickly, using objectively measurable standards whenever possible. Because of their long-term importance to the agreement, and to a successful relationship between the provider and customer, appropriate care should be taken to reach agreement regarding the definition of support services and the standards that will be used to evaluate them. The dynamic tension between the customer and provider in this regard results from the customer "buying a solution" and the provider "selling technology." Usually, the provider will want to use technical metrics (such as size of server, speed of transmission line, etc.) and the customer will want to use operational metrics from its business (such as the ability to process a transaction within so many seconds, or the number of hours from the time a complaint is filed until trouble-shooting by the provider begins).

While "Services and Support" sections typically set forth the affirmative obligations of the provider and the negative covenants of the customer, the model also identifies the customer's responsibilities regarding control of information entered into the provider's systems and access to that system. Model Sections 2.3 and 2.5 hold the customer responsible for its actions. These provisions could be moved to another section of the document dealing with the customer's obligations, but placing them under Article II of the model emphasizes that these risks are controlled by the customer, and that customer is solely responsible for them. Providers may also have specific policies (which they may update regularly to reflect prevailing legal liability) that may be mentioned within these sections (and incorporated into the agreement by reference). Best practice from a customer perspective is to require a provider to provide any such policies as an exhibit to the agreement (rather than a mere reference to a URL) and allow the customer to terminate the agreement should the provider make any unilateral changes that are unacceptable to the customer.

The model also anticipates that a specific number of designated end users will access the services, and that each individual will be provided a unique password (which will be administered by the customer). Customers may seek increased flexibility through provisions that additional end users may be added at any time, provided only that proper payment for any new users is made. Customers may also want to negotiate for a designated number of passwords (simultaneous users) rather than a designated number of unique users (since it is the number of passwords that represents the system capacity that is being priced by the provider, not the number of potential end users). While not included in the model, providers may request a periodic audit of passwords and users to ensure full payment of fees and enforce license rights against end users who are no longer associated with the customer (or, as an alternative, a periodic affidavit from the customer verifying the number of users of the services).

Services and Support

2.1   Provision of Services. Subject to the terms and conditions of the Agreement, Provider will use reasonable commercial efforts to provide the Services to Customer. Customer may request additional services from Provider, but only by submitting a written request to Provider. If Provider accepts Customer's request, then Provider shall provide the additional Services on the terms set forth in the Agreement. Provider will have no obligation to provide any upgrades to the Software. During the term of the Agreement, Provider may make enhancements to the Software and the Services and Customer agrees to use the enhanced versions of the Software and the Services.

A customer-friendly alternative:

> *"Subject to the terms and conditions of the Agreement, Provider will provide to Customer the Services, as more specifically described in Exhibit ___ hereto, in accordance with the Service Levels set forth in Exhibit ___, also attached hereto. Customer may request additional services from Provider, which shall be delivered according to the terms of the Agreement and pursuant to the prices set forth in attached Exhibit ___. During the term of the Agreement, Provider may make enhancements to the Software and the Services at no additional cost to Customer. Customer agrees to use such enhanced versions upon demonstration by Provider that such versions:*
>
> *1.   Will not require new or additional training of Customer personnel, and*
> *2.   Provide functionality equal to or greater than that required by the Agreement or any applicable SOW."*

Note that this alternative language is not without weaknesses. It does not define how the provider is to "demonstrate" that upgrades are acceptable. If the contract is for a large dollar amount, or the services are critical, the better course would be to provide a more substantive/objective definition.

2.2   Grant of License. Subject to the terms and conditions of the Agreement, Provider grants to Customer a limited, non-transferable, non-exclusive license for the term of the Agreement to access via the Internet and use the Services and the Software, but only from the Locations and solely to support Customer's normal course of business.

A more customer-friendly alternative:

> *"...grants to Customer a limited, nontransferable (except as provided in Section ___), nonexclusive license for the term of the Agreement to access via the Internet and use the Services and the Software, provided that Customer's total number of simultaneous users of same shall not exceed ___."*

2.3   Restrictions on Use. Customer may not, directly or indirectly,

(i) License, sublicense, sell, resell, lease, assign, transfer, distribute, or otherwise commercially exploit or make available to any third parties the Services or the Software in any way,

(ii) Alter, modify, translate, or create derivative works based on the Services or the Software,

(iii) Process or permit to be processed the data of any third party,

(iv) Use or permit the use of the Services or the Software in the operation of a service bureau, timesharing arrangement or otherwise for the benefit of a third party,

(v) **Disassemble, decompile, or reverse engineer the Software or any aspect of the Services, or otherwise attempt to derive or construct source code or other trade secrets from the Software,**

(vi) **Build a competitive product or service to the Services or Software or a product or service that uses similar ideas, features, functions, or graphics to the Services or Software,**

(vii) **Use the Services or Software to engage in any prohibited or unlawful activity, or**

(viii) **Permit any third party to do any of the foregoing.**

A customer, particularly a large one, might insist on language similar to:

"Notwithstanding anything to the contrary in this Section 2.3, Customer is hereby authorized to use the Services and the Software for the benefit of Customer's subsidiaries and affiliates at no additional cost or expense to Customer."

2.4   Support Services. Subject to Customer's prompt payment of the fees due under the Agreement, Provider shall provide Customer with the support services specified in Exhibit A. Provider's current support services are set forth in Exhibit B. Customer may upgrade to a higher level of support services without paying a change fee. However, Customer must pay the change fee set forth in Exhibit A to downgrade or cancel support services.

Many customers will seek to remove the last sentence of this section, as paying to *reduce* services is counterintuitive.

2.5   Use of Customer Data/Customer Representations and Warranties. Customer shall be solely responsible for collecting, inputting and updating all Customer Data. Customer represents and warrants that its Customer Information does not and will not include anything that infringes the copyright, patent, trade

secret, trademark or any other intellectual property right of any third party; contains anything that is obscene, defamatory, harassing, offensive, malicious or which constitutes child pornography; or otherwise violates any other right of any third party.

Customer-friendly alternative:

> *"Customer shall be solely responsible for collecting, entering, and updating all Customer Data, and for ensuring that such Customer Data shall not violate any law or regulation, nor infringe the rights, including intellectual property rights, of any third party."*

2.6   **Passwords.**   **Provider shall provide Customer with passwords to access the Service. Customer shall be responsible for all use of its account(s). Customer shall also maintain the confidentiality of all passwords assigned to it. Customer may not share its passwords with third parties or attempt to access the Service without providing a password assigned to it. Passwords cannot be shared or used by more than one individual, but may be reassigned from time to time to new individuals who are replacing former individuals who have terminated employment with Customer or otherwise changed job status or function such they no longer have a need to use the Service or Software. In no event shall Customer allow more individuals to access the Software or Services than the maximum number of passwords provided to Customer under the Agreement and Exhibit A.**

A customer wishing more flexibility might prefer language similar to:

> *"Provider shall provide Customer with passwords to access the Services. Customer may not share its passwords with any third parties. Customer shall be responsible for all use of its accounts. In no event shall Customer allow more individuals to access the Software or Services than the maximum number of passwords provided to Customer under the Agreement and Exhibit ___."*

Consulting Services

One of the benefits of a "master" agreement is that it allows the business relationship of the provider and customer to evolve over time, all within the terms and conditions previously negotiated (preventing surprises resulting from changes in the business environment, management contacts, etc.). This evolution is typically documented in SOWs attached to the master agreement as the need arises. Each SOW defines a discrete set of deliverables and timelines at a very technical level, all of which are generally governed by the master terms and conditions.

The model agreement provides that SOWs are to be separately negotiated and executed by both parties prior to becoming effective. It specifically contemplates that the relationship will morph over time, as the needs of the parties require.

The model explicitly provides that its terms and conditions will prevail over any SOW. However, the parties may wish to permit individual SOWs to change the basic terms of the agreement. Such a provision might read:

> *"In the event of a conflict between the terms of this Agreement and any SOW, the terms of this Agreement shall control, unless the SOW in question specifically and explicitly modifies the terms of this Agreement. Any such modification, however, shall apply only to the SOW in which it appears."*

3.1   Statements of Work. Upon Customer's request and subject to both parties' acceptance of a Statement of Work, Provider may modify, customize or enhance the Software or provide implementation, data transfer, training, or other services relating to the Software or Services ("Consulting Services"), for an additional fee. Customer shall request Consulting Services by completing a Provider order form, in the form attached as Exhibit C. Upon receipt of Customer's written request for Consulting Services, Provider shall prepare a Statement of Work. A Statement of Work is an offer to perform Consulting Services on specified terms and for specified fees. A Statement

of Work will only be binding if signed by both parties. Each Statement of Work will be governed by the Agreement. In the event of a conflict between a Statement of Work and the Agreement, the terms and conditions of the Agreement will prevail.

Since the specifics of the consulting services required over the term of the agreement may not be known as of the date of the master agreement, the agreement provides a mechanism for appropriately pricing such services. The model provides two alternative pricing options (one of which must be explicitly set out in an SOW):

1. "Fixed fee" or
2. "Time and materials."

With respect to time and materials pricing, best practice is to include an initial rate card of standard pricing for the type of personnel/services that may be anticipated. The rate card may be amended over time by agreement of the parties. Similarly, some benchmark should be applied to expenses. The model refers to "reasonable" expenses. However, sophisticated customers will want to (i) include specific line item expense policies (e.g., "flight expenses to be reimbursed as coach rates" or "hotel expenses not to be in excess of [$___] per night") and/or (ii) require prior written approval from the customer before the provider incurs any expense in excess of the limits set forth in the agreement or applicable SOW.

Administrative expenses may also be negotiated or defined with more certainty between sophisticated parties. In all cases, it is in both parties' best interest to separately itemize expense items.

Customer-friendly alternative:

> *"Customer may, from time to time, request additional services from Provider by submitting an order form, in the form attached as Exhibit ___. Such additional services shall be priced pursuant to Exhibit ___. Upon receipt of Customer's order form, Provider shall prepare a*

Statement of Work, which shall detail the services to be provided, applicable service levels and all fees and expenses. Such a Statement of Work shall become binding when signed by both parties to this Agreement."

3.2   Payment for Services. Unless a Statement of Work expressly states that Consulting Services are to be performed for a fixed fee, Provider shall provide Consulting Services to Customer on a "time and materials basis" with any price quotation constituting merely an estimate that is not a binding on Provider. For purposes of this Agreement, "time and materials basis" means that Customer shall pay Provider at the hourly rates specified by Provider for the Consulting Services and shall reimburse Provider for expenses as set forth in section 3.3.

Model Section 3.2 is provider-friendly, as it allows the provider wide pricing latitude. Customers would typically seek to delete Section 3.2 and change the "default" to fixed fee pricing. Even when fixed pricing is not practicable, astute customers will seek to control their total exposure. For example, a customer may require that an estimate becomes binding when the SOW is signed, or require that cost "shall not exceed" the estimate without the customer's prior written consent. Providing cost estimates is not an exact science, and a provider may genuinely need some flexibility. At the same time, no customer wishes to sign the proverbial "blank check."

3.3   Expenses. In addition to payment for the Consulting Services, Customer shall reimburse Provider for reasonable travel, administrative, equipment, and out-of-pocket expenses incurred in performing the Services, in accordance with section 6. These expenses are not included in any estimate in a Statement of Work unless expressly itemized.

"Reasonable expenses" is another term that invites dispute because it has no generally accepted meaning. Is "reasonable" airfare first class or coach? Large customers tend to require providers to adhere to the customer's own internal travel standards and policies. Smaller customers might agree to reimburse expenses, but only to a set dollar limit.

Termination

Terminating the master agreement is the ultimate recourse for both parties, and it requires a complete accounting between the parties. The model provides for automatic termination of all SOWs in the event that the master agreement is terminated. Conceptually, this makes sense (in that each SOW is dependant upon the terms and conditions of the master agreement). However, there may be instances in which "winding down" or "transition services" are required with respect to specific SOWs. From a practical perspective, best practice is to provide that all SOWs <u>may</u> be terminated if only one SOW is terminated, or upon notice that the master agreement is to be terminated, allowing both parties to address the various types of services being performed under each SOW.

The model is also provider-friendly in that it ends most obligations of the provider upon termination of the agreement, requires immediate payment of all fees incurred prior to the date of termination, and provides for "wind down" expenses to be paid to the provider. Obviously, some obligations of the provider should be ongoing (and specific sections, such as confidentiality, should explicitly state that they will survive termination of the agreement).

On the other side of the table, the customer may seek provisions that:

- State that the provider is due payment only for undisputed fees for services performed prior to termination
- Require some form of transition services
- Require the provider to continue to preserve the confidentiality of customer data (provider will also want continued protection for its own intellectual property and confidential information).

Termination is generally a last resort, and best practice teaches that drafters should incorporate a process requiring formal notice of dispute and creating a procedure to ensure that such disputes will be reviewed by appropriate levels of management for both sides before termination may be invoked. Indeed, the more vital the services are to the customer, the more difficult the customer may want to make the standard for termination. Such a termination provision might read:

> *"The parties hereby agree that neither shall be entitled to terminate this Agreement for any perceived or alleged breach or default unless:*
>
> - *Written notice of said default or breach is given, pursuant to Section ___ hereof; and*
> - *Said default remains uncured for more than 30 days after the date of said notice; and,*
> - *The party in breach or default does not, within 30 days of the date of notice submit a written request for conference to permit the appropriate corporate officers to resolve the matter, said conference to occur within 14 days of the date of such request, or at the earliest date the parties can otherwise arrange, agreeing that such conference shall be given a priority.*
>
> *It is further agreed that the parties attending or participating in such conference shall be vested by their principals with all authority necessary to resolve the matter at hand without reference to higher authority."*

3.4 Effect of Termination during Performance of Consulting Services.

Upon termination of the Agreement, Provider will be relieved of all obligations to perform Consulting Services, the Statement(s) of Work for all Consulting Services will terminate, and Customer immediately shall pay Provider for Consulting Services performed prior to the date of termination. Should Customer terminate any Statement of Work prior to the conclusion of the Statement of Work, Customer shall reimburse Provider for any costs associated with winding down Consulting Services under that terminated Statement of Work.

Customer-friendly alternative:

> *"Upon termination of this Agreement:*
>
> •   *Provider shall be relieved of all obligations to provide Services under the Agreement and any SOWs then in effect, but Provider shall immediately begin to provide Termination Services, if any, as may be specified in this Agreement or any applicable SOW.*
>
> •   *Customer shall promptly pay Provider undisputed fees due for Services performed prior to the effective date of Termination and shall pay for Termination Services as required by this Agreement or any applicable SOW."*

The model is provider-friendly with respect to non-monetary defaults, providing for notice of breach and an extended time period in which to cure. Such notice and cure is not provided to the customer for monetary breaches.

This section of the agreement tends to be tempered by the negotiating strength of the parties. Larger customers contracting for business-critical functionality will almost never allow a provider to terminate the agreement or a statement of work for a simple monetary default. Most large customers will insist upon notice and a chance to cure monetary defaults that mirrors the notice and cure provisions granted to providers with respect to non-monetary default. Some very large customers with significant negotiating strength will flatly prohibit termination by the provider for non-payment (insisting that the risk of bringing the customer's business to a halt as a result of a payment dispute is much larger than any risk associated with ultimate recovery of fees from the customer, even if such recovery eventually involves litigation).

This section of the agreement is also typically negotiated based upon actual default scenarios (such as a failure of the provider to perform according to SLAs or customer's failure to pay on time). However, even when both parties are acting in good faith, a default is sometimes difficult to distinguish from a misunderstanding. The classic illustration would be a receivables processing term of thirty days by the provider and a payables processing cycle of forty-five days by the customer, the end result being that the administrative systems of both parties can create conflict, even when both parties attempt to perform. Many times, drafters are not aware of their client's internal limitations in this regard, creating technical defaults that will become a source of potential conflict over time.

In some instances, the agreement or the individual SOWs may also provide for cross-defaults, such that a termination of one SOW will be sufficient to terminate other named SOWs or the agreement itself.

3.5   Termination of Statement of Work/Right to Cure.

(i)   Either party may terminate the Statement of Work if the other party breaches a Statement of Work and has not cured its breach within 30 days of written notice from the non-breaching party. In addition, Provider may terminate the entire Agreement if Customer breaches any payment obligation. Otherwise, however, neither party may terminate the entire Agreement based solely on breach of a Statement of Work.

(ii)   Consent to extend the cure period for breaches of a Statement of Work may not be unreasonably withheld so long as the breaching party has in good faith begun to cure its defects in performance during the 30 day notice period. Notwithstanding the foregoing, Provider, in its sole discretion, may withhold its consent to extend the cure period for Customer's breach of its payment obligations.

A more customer-friendly version of model 3.5(i) would omit the second sentence, or change it to permit termination only after repeated failure to pay. A customer-friendly version of model 3.5(ii) would omit the last sentence, or at least defer termination until the occurrence of several late or missed payments.

Change Orders

In any contract to be performed over time, it may become necessary to change the nature or scope of the services to be provided. Circumstances may change or the parties may discover that their initial assumptions were inaccurate. Therefore, it may be appropriate to change the project from time to time. However, it is important to implement change in an orderly fashion. Neither party is well served if the other party launches new and expensive initiatives, only to discover that the customer cannot afford the bill, the provider can't deliver as promised, or senior management of either party disavows the new initiative.

As a result, it is typical for master agreements, such as the model, to specifically state who may request and approve changes and to detail how the changes will be negotiated and documented. The language below, model Section 3.6, is slightly provider-friendly in that it requires the provider to provide a fee estimate, rather than a fixed price. Note also that this language contemplates that only the customer will propose changes. Arguably, such a provision will protect the customer from a provider's efforts to sell additional features and costs, but it could prove an obstacle in those instances where provider has identified a genuine need for change. An alternative approach would be to require that any changes be proposed by and to appropriate levels of project management for each of the parties, to ensure that both parties receive and properly review changes (rather than focusing on who has the prerogative of proposing a change).

> **3.6   Change Orders.** Provider's obligation to perform Consulting Services is limited to the services described in a Statement of Work. Customer may request changes to a Statement of Work by submitting proposed changes in writing to Provider. Within 14 days of receiving Customer's request, Provider will provide Customer with a proposed change order. A change order will set forth, at a minimum, (i) a written description of the changes to the Statement of Work, (ii) any changes to the services schedule, and (iii) any changes to the fee estimate. A change order will be binding only if signed by both parties, in which case it will be governed by the terms and conditions of the Agreement. In the event of a conflict between a change order and a Statement of Work, the Statement of Work will prevail.

Cooperation

Complex transactions generally entail significant work by personnel of both parties and can require considerable cooperation and coordination. For example, provider personnel may require physical access to the customer's data center to install specialized hardware. If the customer is not prepared to allow the installer sufficient access to accomplish this task, the project will likely be delayed. As a result, agreements such as the model, tend to include "cooperation" clauses in which each side pledges to assist the other as needed. More than that, such clauses can excuse performance by one side if the other fails to perform its assigned tasks (i.e., where performance by one party is conditioned upon an act of the other). Note that model Section 3.7 is a rather provider-friendly example. It provides that the customer will provide assistance to the provider and excuses the provider's performance if that customer assistance is not forthcoming.

A wary customer might seek to delete the first sentence of Section 3.7 or provide defined, objective limits on the nature and extent of the cooperation required. A very wary customer might delete the sentence and add to the end of the paragraph language along these lines:

> *"Notwithstanding any provisions to the contrary in this Agreement, Provider shall be solely responsible for the complete, accurate, and timely performance of the Services, and no delay in performance shall be excused by reason of Customer's failure to act or provide assistance unless such obligation is specifically assigned to Customer in this Agreement or an applicable SOW."*

3.7   Cooperation. Customer shall cooperate with and assist Provider in performing Consulting Services. Among other things, Customer shall provide to Provider timely access to office accommodations, facilities, and equipment; complete and accurate information and data from its officers, agents, and employees; and suitably configured computer products. Customer's failure to do so will relieve Provider of responsibility for any related deficiencies in its performance.

Customer Restrictions and Responsibilities

Given the ever-increasing sensitivity to data security and data privacy, it is common for providers to require that customers commit to use the provider's systems only in a lawful manner, and only to process information the customer is lawfully entitled to possess and process. Model Section 4.1 is illustrative of such provisions. It is unclear, however, what remedies and what protections model Section 4.1 would give the provider in the event that the customer were to use the provider's systems inappropriately. On the other hand, the customer might be better served by straightforward prohibitions such as:

> *"Customer shall use the Services only in accordance with the Documentation and in accordance with applicable laws and regulation."*

Care should be taken to define "documentation" in such a way that the provider may not reduce the customer's rights by making unilateral changes to the underlying materials (e.g., by changing the materials posted on provider's Web site.)

4.1 Compliance with Documentation and Laws/Representations and Warranties. Customer represents, covenants, and warrants that Customer will use the Services and any data of third parties only as contemplated by the Documentation and in compliance with all applicable laws and policies (including but not limited to laws, government regulations, Provider policies and any other applicable policies relating to intellectual property, employment, labor, spamming, spoofing, network security, privacy, obscenity or defamation).

SaaS providers and, to a lesser extent, ASP providers, make money providing a standardized service. Ensuring that an individual customer's infrastructure is sufficient to run those services is generally outside the provider's business model. As a result, ASP and SaaS agreements typically contain provisions such as model Sections 4.2 and 4.3, making it clear that the customer is responsible for providing adequate infrastructure and for making sure the provider's services truly match the customer's needs. In the event the provider is being engaged to provide consultation regarding infrastructure or suitability, care should be taken to remove these provider-friendly disclaimers from the model.

4.2   Customer Equipment. Customer shall be responsible for selecting, obtaining and maintaining any equipment and ancillary services needed to connect to, access or otherwise use the Services, including, without limitation, modems, hardware, servers, software, operating systems, networking equipment, web servers, and long distance and local telephone service, but excluding the Software (collectively "Equipment"). Customer shall be responsible for ensuring that the Equipment is compatible with the Services and the Software and complies with the Documentation. Customer shall also be responsible for the security and use of the Equipment.

4.3   Customer Requirements. Customer shall be responsible for determining what Services, Software and Consulting Services and support it requires.

Confidentiality and Publicity

Control of confidential information and publicity can have a significant impact on a customer's fortunes. Thus a careful draft will address both in some detail.

At a theoretical level, confidential information can be divided into two broad categories: information protected for economic reasons (e.g., information that provides a competitive advantage) and information protected for legal reasons (e.g., personal health information protected by the Health Insurance Portability and Accountability Act). Drafters should take care to determine whether a given transaction will involve legally protected information and, if it does, ensure that proper protections are in place.

With regard to "economic information," best practice is to avoid defining confidential information as "all information" one party might provide to the other. Such a definition gives comfort to the disclosing party, but it imposes a potentially large burden on the recipient, who could be asked to protect everything from valuable trade secrets to routine lists of the discloser's locations. In that event, the disclosing party runs the practical risk that, rather than protecting everything, the recipient will protect nothing. Better practice is to narrowly define confidential information and require that it be marked or identified as confidential at the time of disclosure.

A well-drawn confidentiality provision will also specify how the confidential information may and may not be used, and when it will be returned or destroyed.

Model Section 5.1 is representative of commonly seen provisions:

- The term "confidential information" is broadly defined and includes the terms of the contract itself.
- Confidential information is to be used only for the performance of the contract and may be disclosed only to employees and agents who owe their principal a duty of confidentiality.
- Confidential information is to be returned upon termination of the agreement.
- Should any law or regulation require a more stringent standard, that standard shall apply.

Model Section 5.2 sets out standard exceptions to the definition of confidential information and directs how the receiving party will respond to legal orders to release confidential information of the other.

In the context of ASP/SaaS transactions, the potential exposures of the provider and customer with respect to disclosure of proprietary information may be quite different. Unless the customer receives the provider's software, the customer may receive little or no confidential information from the provider. On the other hand, the provider may receive access to some of the customer's most sensitive data, such as customer lists, pricing and price lists, patient data, or personal financial information. Thus the customer's drafter, in particular, should take care to avoid agreeing to protect confidential information from the provider (or using reciprocal clauses) if little information of value will really be received.

5.1   Confidentiality Obligation. "Confidential Information" means all written or oral information designated as confidential at the time of disclosure that is made accessible to the other party in connection with the Agreement including, without limitation, computer programs, software, formulas, data, information, inventions, techniques, strategies, trade secrets, know-how, plans for products or services, marketing plans, financial documents or data, processes and designs, Service passwords, and the terms, but not the existence of, the Agreement. Written Confidential Information must be marked as "confidential" or "proprietary." Oral Confidential Information must be designated as confidential at the time of disclosure and reduced to a written summary and marked "confidential" or "proprietary" within 10 days of the oral disclosure. Each of the parties shall treat the other party's Confidential Information confidentially and with at least the same degree of care it uses to prevent the disclosure of its own Confidential Information, but in no event less than reasonable care. In addition, each party shall use the Confidential Information of the other party solely in the performance of its obligations under the Agreement and not disclose it, except to authorized employees of the receiving party or its affiliates, its legal counsel and its accountants (provided that the receiving party contractually obligates them to a duty of confidentiality no less restrictive that the duty imposed by this section 5.1 and remains jointly and severally liable for any breach of confidentiality by them). Each party shall promptly notify the other party of any actual or suspected misuse or unauthorized disclosure of its Confidential Information. Upon expiration or termination of the Agreement, each party shall return all tangible copies of any Confidential Information received from the other party. To the extent that relevant federal or state laws or regulations may specify a higher level of confidentiality than set out in the Agreement with respect to certain forms of Confidential Information, such higher standard shall govern as between the parties.

5.2   Exclusions. Confidential Information will not include information that the recipient can prove:

(i) was generally available to the public at the time it was disclosed,

(ii) was known to the recipient, without restriction, at the time of disclosure by the disclosing party,

(iii) is disclosed with the prior written approval of the disclosing party,

(iv) was independently obtained or developed by the recipient without any use of the Confidential Information,

(v) becomes known to the recipient, without restriction, from a source other than the disclosing party who does not owe a duty of confidentiality to the disclosing party and obtained the information by lawful means, or

(vi) is disclosed in response to an order or requirement of a court, administrative agency, or other governmental body, a subpoena, or by the rules of a securities market or exchange on which the disclosing party's securities are traded (but only if (a) the recipient vigorously opposes the proposed disclosure and provides prompt advance notice to the disclosing party to enable it to appear and independently contest the disclosure, and (b) any Confidential Information so disclosed will otherwise remain subject to the provisions of this section 5). The burden of proof in establishing that any Confidential Information is subject to any of the foregoing exceptions will be borne by the receiving party.

Trademark License

Model Section 5.3 addresses a standard provider request: "May we use your name as one of our clients?" The answer given by Section 5.3 is very provider-friendly, for it grants broad permission to use the customer's name, logo, and trademarks. This broad language could give customers pause for a number of reasons:

- Some companies simply do not allow others to use their names or marks. Others will allow such use provided:

    o They receive consideration in return for the permission.
    o They are given prior approval over the announcement or advertisement (generally, to ensure that the publicity cannot be taken as an endorsement of the provider's products or services.)

- It contains no requirement that the provider's use conform to the customer's own use standards. Thus a careless provider could change the color, size, or content of a registered mark or logo and, as a result, endanger the time and expense the customer has invested to turn that mark into a valuable commercial asset.

A more customer-friendly provision might read:

> *"Customer will consider requests for publicity rights on a case by case basis, but may deny such requests solely in Customer's sole discretion. No portion of this Agreement shall be deemed to grant to Provider any right or license to the names, logo and trade or service marks of Customer."*

5.3   Trademark License. Customer grants Provider a limited, nonexclusive, worldwide right and license during the Term to use the names, logos and trademarks of Customer to publicize the existence of the business relationship established by the Agreement.

Payment of Fees

In most ASP/SaaS agreements, the customer's principal affirmative obligation is to pay (and, secondarily, to pay on time).

Model Section 6.1 is provider-friendly in that it requires the customer to pay and permits the provider to change prices unilaterally. The requirement that the customer render payment within thirty days of the date of invoice may also be an invitation to dispute. Many customers, particularly large ones, may have their accounts payable systems designed to issue payment within forty-five days of receipt of invoice.

A more customer-friendly provision might be:

> *"Customer shall pay to Provider the fees for Services as set forth in Exhibit ___. Payment shall be due within ___ days of Customer's receipt of invoice."*

6.1 <u>**Fees.**</u> **Customer shall pay Provider the fees for the Services as set forth on Exhibit A. Provider reserves the right to change the fees or applicable charges upon 30 days prior notice by e-mail or otherwise. Unless otherwise specified in Exhibit A, Customer shall pay the fees set forth in the invoice to Provider within 30 days of the invoice date.**

Even parties acting in the utmost good faith may from time to time disagree. Best practice teaches the need to provide dispute resolution provisions, whether the dispute focuses on the quality of services delivered or the accuracy of an invoice.

Model Section 6.2 is quite provider-friendly, as it requires the customer to provide written documentation of any disputed balances before the provider will be required to provide credit or adjustment. Implicit in this language is a prohibition: the customer is not given the option of withholding payment of disputed balances.

Thus, a customer-friendly alternative might be:

> *"Customer may withhold payment of any disputed balance, provided Customer:*

- *Shall pay any undisputed balance when due;*
- *Promptly provide written notice of the nature of the dispute and, if requested by Provider, enter into discussions to resolve the dispute pursuant to Section ___.*

Subject to the foregoing, withholding payment of disputed balances:

- *Shall not constitute a breach of this Agreement;*
- *Shall not permit Provider to delay, diminish, interrupt or in any other way reduce the timely performance of the Services."*

6.2   Disputed Amounts. Provider need not provide any adjustments or credits to Customer for disputed amounts billed by Provider unless Customer provides written notice of the disputed amounts to Provider within 60 days after the invoice date on the first invoice in which the error or problem appeared. Customer must direct any notices required under this section 6.2 (and any related inquiries) to Provider's customer support department.

"Time is money," particularly if a large invoice is unpaid and potential investment income from the expected revenue is being lost (or isn't booked in the fiscal period where it will have the most positive effect for the provider). As a result, providers typically seek late payment provisions to ensure the time value of money expected from the agreement. Model 6.3 is a provider-friendly example of such a provision. Customers would typically press for the lowest possible interest rate and reject the language regarding obligation for the provider's collection expenses.

6.3   Interest and Taxes. Unpaid fees are subject to a finance charge of [insert interest rate] percent per month on any outstanding balance, or the maximum permitted by law, whichever is lower, plus all expenses of collection (including reasonable attorneys' fees).

Term and Termination

"Term and Termination" provisions address a number of fundamental questions:

- How long will the agreement last?
- Under what circumstances may the agreement be dissolved before its agreed date?
- What will happen when the agreement ends?
- May the agreement be extended and, if so, for how long and at what price?

Note that some drafters distinguish between expiration and termination, using "expiration" to denote dissolution of the contract on its agreed end date and "termination" for any end before that agreed date. The distinction has a certain attraction: implicit in "termination" is the suggestion that the end results either from dispute or changed circumstances. It also permits another useful distinction:

- "Termination services," being services required after a dispute;
- "Transition services," being services required after an agreed or automatic end to the agreement.

The "term" of the agreement defines how long the agreement will be in force—how long the provider will provide the services and for how long the customer will pay the fees. In the context of a master agreement, the precise meaning can be complicated by the addition of SOWs. In that event, certain questions will inevitably arise. Does "term" mean the duration of the master agreement or the master plus the last of the SOWs? Will terminating an individual SOW terminate the master? Will the SOWs survive termination of the master?

Model Section 7.1 is rather standard, but note the last sentence: renewal requires a mutual agreement. In other words, any renewal could require extensive new negotiations before mutual agreement is reached. The customer might prefer:

> *"Customer may elect to renew this Agreement for up to __ additional terms of __ years each, upon __ days advance written notice to Provider. Pricing for each renewal term shall not increase by more than __ percent"*

Model Section 7.2(i) is essentially a buyout provision. The customer may stop using the services, but must still pay for them for the remainder of the term. Such language could provide the provider a windfall, as the provider could be paid for overhead expenses it will no longer incur. As a result, customers can be expected to press for a reduced buyout price.

Model Section 7.2(ii) has a standard structure, but as discussed earlier, requiring a customer to respond to payment questions within fifteen days may simply not be administratively practical (especially for a large, multinational customer). In addition, the reference to "material breach" may create some ambiguity in its attempts to create flexibility. Like "commercially reasonable effort," there is no accepted, objective standard for what is or is not a material breach. Separate clauses can be added that provide objective measures of "materiality" (such as "failure of Customer to pay fees when due for more than __ consecutive months" or "failure of Provider to meet the Service Levels for more than __ consecutive months or more than __ months out of any twelve consecutive months").

Model Section 7.2(iii) is also relatively standard, but it may create issues with respect to the automatic stay provisions of the Bankruptcy Code. The intent is to allow termination upon notice of these financially distressing situations, rather than having to wait until the financial distress affects either party's ability to perform.

7.1   Term. Unless terminated earlier in accordance with section 7.2, the term of the Agreement will be [insert term] years. The Agreement may be renewed by mutual written agreement of the parties.

7.2   Termination. The Agreement may be terminated as follows:

(i) Customer may terminate the Agreement at any time by notifying Provider in writing and paying all undisputed fees for the Services for the remainder of the then-current term.

(ii) Either party may terminate the Agreement immediately by written notice if the other party materially breaches the Agreement and fails to cure its breach after receipt of written notice within **(a)** 15 days in the case of non-payment of any fees, or **(b)** 30 days in the case of all other breaches.

(iii) Either party may terminate the Agreement immediately by written notice if the other party **(a)** becomes insolvent, **(b)** makes an assignment for the benefit of creditors, **(c)** files or has filed against it a petition in bankruptcy or seeking reorganization, **(d)** has a receiver appointed, or **(e)** institutes a proceeding for liquidation or winding up. In the case of involuntary proceedings, a party will only be in breach if the applicable petition or proceeding has not been dismissed within 90 days.

7.3 **Effect of Termination.** Within 7 days of expiration or earlier termination of the Agreement, Customer shall pay to Provider all undisputed fees for the Services up to and including the date of termination.

Regardless of the termination provisions included within the agreement, certain obligations should extend beyond the end of the agreement. For example, provisions with respect to maintenance of confidential information must continue (to the full term of confidentiality agreed to between the parties). From a drafting perspective, this section is usually reserved until all substantive negotiation is concluded and the specific section references to clauses that will survive are then conformed.

7.4 **Survival.** The provisions of sections 2.3, 2.5, 2.6, 3.2, 3.3, 3.4, 4.1, 5.1, 5.2 and 6 though 12 will survive the expiration or earlier termination of the Agreement. Customer's obligations under section 5 with respect to the Software and Services shall survive the expiration or earlier termination of the Agreement for a period of 2 years.

Proprietary Rights

Providers take particular care to protect their proprietary information—their software, know-how, business processes, system architecture, and the like are typically major assets of their business. Thus customers should expect language protecting the intellectual property and proprietary interests of the provider. But that does not mean the provider language must be, or should be, accepted automatically. The customer will want to be sure it receives the use rights it reasonably anticipates needing (including potential access to source code—see the discussion above regarding escrow). The customer may also need the right to modify the provider's applications (this is more likely to occur in the context of an ASP agreement than an SaaS transaction), in which case the customer's representatives would need to bargain for, and secure, the right to create derivative works from the provider's materials. In this regard, the section should also explicitly define the rights of both parties to those derivatives.

It is also possible that the provider will be given access to some of the customer's proprietary information, in which case the agreement must provide an appropriate license for the provider to use such information in providing services while at the same time giving the customer adequate protection. Because proprietary rights are potentially of great commercial value, this subject should be reviewed with great care.

Model Section 8.1 is a representative, provider-friendly provision:

- The provider retains ownership of its proprietary materials.
- The provider may place, and the customer shall not remove, copyright notices.
- The provider may use methods or techniques it learns from this project or customer for other projects and customers.

The last point, regarding what the provider may learn from a given project, refers to "residuals," which deserve special attention. During the course of a project, the provider's personnel are likely to learn about how the customer does business and new ways to deliver the provider's services (or information technology services in general). But human memories, unlike computer hard drives, cannot be erased and reformatted. Thus residual provisions attempt to limit use of retained knowledge to that which is intangible and is not commercially valuable to the customer. In the context of an SaaS transaction, residuals may be a slight concern. The provider personnel may never have access to the customer's truly valuable information or processes. The concern rises, however, in direct proportion to the provider's access to the customer's business. If, for example, the provider is engaged to review, redesign, implement, and host the customer's new financial data warehouse, the question of residuals would require careful consideration.

The model agreement proprietary rights provision is rather one-sided. It protects the rights of the provider, but not of the customer. A customer-friendly alternative provision might be:

> *"Provider's Proprietary or Confidential Information shall remain the sole and exclusive property of Provider. Customer's Proprietary or Confidential Information shall remain the sole and exclusive property of Customer. Neither party shall have any interest in, nor any right to use (including, without limitation, any use resulting in disclosure to any third party) the other parties' Proprietary or Confidential Information except as specifically provided for by the Agreement or as otherwise permitted and specified by separate written agreement executed by both parties hereto."*

8.1   Provider's Proprietary Rights. Exclusive of Customer Information, Provider (or its third-party licensors, if applicable) will retain all rights, title, and interest in and to the Software, Services, and the Provider Information and all legally protect able elements or derivative works thereof. Provider may place copyright and/or proprietary notices, including hypertext links, within the Services. Customer may not alter or remove these notices without Provider's written permission. Customer may not have the right to, and agrees not to, attempt to restrain Provider from using any skills or knowledge of a general nature acquired during the course of providing the Services, including information publicly known or available or that could reasonably be acquired in similar work performed for another Provider customer.

8.2   Customer's Proprietary Rights. Customer will retain all rights, title and interest in and to the legally protect able elements of Customer Information and derivative works thereof.

Indemnity

A thorough contract allocates, to the extent possible, both responsibilities and risks. Indemnification provisions are part of the risk allocation process. They are intended to protect one side from liability or expense arising from the conduct of the other. For example, the provider might want to protect against loss if the customer uses data or information to which the customer does not have proper rights. On the other hand, the customer would want to be protected against claims that the provider's software infringes the rights of third parties, or that the provider has breached the privacy rights of the customer's clients.

The model is provider-friendly. The provider is protected against virtually any claim that might arise from the customer's conduct, while the customer is given no protection.

Customer will indemnify, defend, and hold Provider harmless from and against any and all costs, liabilities, losses and expenses (including, but not limited to, reasonable attorneys' fees) resulting from any claim, suit, action or proceeding brought by any third party against Provider arising out of or relating to Customer's breach of its representations or warranties hereunder or its use of the Software and Services.

An alternative provision more palatable to the customer might be:

"At Providers expense as provided herein, Provider agrees to defend, indemnify, and hold harmless Customer, its Users, directors, officers, agents, employees, members, subsidiaries, joint venture partners, and predecessors and successors in interest from and against any claim, action, proceeding, liability, loss, damage, cost, or expense (including, without limitation, attorneys' fees as provided herein) arising out of

any alleged act or failure to act by Provider or its directors, officers, agents, subcontractors, or employees, (including, without limitation, negligent or willful misconduct) alleged to:

1.   Cause any injury to any person or persons or damage to tangible or intangible property,

2.   Breach the provisions of Section ___ relating to Providers use of confidential information owned or controlled by Customer, or

3.   Have breached the Agreement (collectively referred to for purposes of this Section ___ as "Claim(s)").

Provider shall pay all amounts that a court finally awards or that Provider agrees to in settlement of any Claim(s) as well as any and all reasonable expenses or charges as they are incurred by Customer or any other party indemnified under this Section ___ in cooperating in the defense of any Claim(s).

At Customer's expense as described herein, Customer agrees to defend, indemnify, and hold harmless Provider, its agents and employees, from and against any claim, action, proceeding, liability, loss, damage, cost or expense (including, without limitation, attorneys' fees as provided herein) arising out of any alleged act or failure to act by Customer or its directors, officers, agents or employees (including, without limitation, negligent or willful misconduct) alleged to cause any injury to any person or persons or damage to tangible or intangible property (collectively referred to for purposes of this section ___ as "Claim(s)"). Customer shall pay all amounts that the court finally awards, or that Customer agrees to in settlement of any Claim(s), as well as any and all expenses or charges as they are incurred by Provider in cooperating in the defense of any Claim(s)."

Warranty and Disclaimer

In general terms, a warranty is a commitment from one party to the other that "X is a fact" or that "Y will be performed according to 123." Because warranties are generally given only for material provisions, breach of a warranty can be grounds for termination of the agreement.

While model Section 10.1 uses the common standard "reasonable commercial efforts," most warranty provisions should create measurable, objective standards. Similarly, model Section 10.1 provides for only one remedy: termination of the contract. More expansive sections would provide alternatives that escalate remedies based upon the actual damage the specific type of warranty breach would occasion.

Model Section 10.1 is also provider-friendly. The warranty applies only so long as the customer:

- is in compliance with the documentation,
- has paid fees when and as due, and
- has made no unauthorized changes in the software or service.

The first point, regarding compliance with documentation, has some merit, but the customer would want to guard against unilateral changes in the documentation. The second point is misplaced. Payment and disputed payments should be dealt with separately from warranty provisions. (And disputed payment situations should not give rise to loss of negotiated warranty protections.) The last point is quite reasonable. The provider may not have the time or expertise to deal with problems caused by the customer's own tinkering (or combinations the customer can create which the provider never anticipated).

10.1 <u>Limited Warranty for Services</u>. Provider shall use reasonable commercial efforts consistent with prevailing industry standards to maintain the security of the Services and minimize errors and interruptions in the Services, provided that:

(i)   Customer uses the Service and the Software strictly in accordance with the Documentation,

(ii)   Customer pays all amounts due under the Agreement and is not in default of any provision of the Agreement, and

(iii) Customer makes no changes (nor permits any changes to be made other than by or with the express approval of Provider) to the Software or Service.

In addition, Customer acknowledges that the Services may be temporarily unavailable for scheduled maintenance, for unscheduled emergency maintenance, or because of other causes beyond Provider's reasonable control (including without limitation delays or other problems inherent in the user of the Internet and electronic communications services). Provider will not be liable to Customer as a result of these temporary service interruptions.

A more customer-friendly alternative might be:

> *"Provider represents, warrants, and agrees that during the Term of the Agreement, the Provider System and all Component(s) shall perform in accordance with the Specifications."*

Note that this language provides an express standard: "in accordance with the Specifications."

While we have previously discussed the issues arising from non-objective terms such as "accepted industry practices," model Section 10.2 (below) uses this subjective term. Preferred wording might make reference to published specifications. Similarly, model Section 10.2 provides that the provider is to re-perform the services and, if the services are still deficient, refund to the customer the fees paid. However, model Section 10.2 does not specify how promptly the provider will re-perform, nor does it provide clear standards regarding what will constitute "acceptable" performance. Moreover, a cash refund might be a totally inadequate remedy if breach of this warranty leaves the customer without a solution to a pressing or important business need.

10.2 Limited Warranty for Consulting Services. Provider warrants to Customer that any Consulting Services will be performed in a manner consistent with generally accepted

industry practices. In order to receive this warranty, Customer must report any deficiencies in the Consulting Services to Provider in writing within 90 days of completion of the services identified in a particular Statement of Work. Any deficiencies not reported within such 90 day period shall be deemed waived by Customer.

Disclaimers of Warranties

Disclaimers of warranties are a standard part of all contracts, if only because statutory law imposes certain warranties unless they are explicitly disclaimed. One disclaimer frequently encountered is the disclaimer of warranty of fitness for a particular purpose. In most cases, a provider will not know the specific needs or circumstances of customer. In such cases, it is entirely appropriate that a provider would not warranty it goods or services to fit those needs. However, in some situations, the provider will be "the expert." In such circumstances, it might be appropriate for the customer to rely upon the provider to determine the "fit" between the provider's solution and the customer's needs (making the disclaimer of fitness for a specific purpose inappropriate).

Software agreements also generally disclaim any warranty that operation of the software will be uninterrupted. After all, software is complex and prepared by humans and may fail despite the provider's very best efforts. Model Section 10.3 reflects this logic. The provider disclaims any warranty that the services will be uninterrupted, any warranty regarding the quality of the output, any warranty regarding the quality of the services, and any warranty that the provider's offerings do not infringe the rights of any third party.

From the perspective of the customer, a better provision might be:

> *"Provider hereby disclaims all other warranties express or implied, including, but not limited to, implied warranties of merchantability or fitness for a particular purpose."*

In this later customer-friendly version, all other points discussed in model Section 10.3, such as quality of consulting or interruption of service, should be addressed through the service levels and the credits or other remedies available if service levels are not met. Such language protects provider against implied or statutory warranties that may not apply to the transaction at hand while allowing the customer to use service levels and service credits for protection against defects in the products or services.

Note that in most states disclaimers of warranties must be sufficiently distinctive from the surrounding text so as to stand out (and in some states, the minimum font or point size is a statutory mandate). Practically, this requirement usually means the drafter must use all caps, boldface, larger type size, or some other formatting characteristic.

10.3 DISCLAIMER OF WARRANTIES FOR SERVICE AND SOFTWARE. NEITHER PROVIDER NOR ITS SUPPLIERS OR SERVICE PROVIDERS WARRANT THAT THE SERVICES WILL BE UNINTERRUPTED OR ERROR FREE, NOR DO THEY MAKE ANY WARRANTY ABOUT THE RESULTS THAT MAY BE OBTAINED BY USING THE SOFTWARE OR SERVICES. EXCEPT AS EXPRESSLY AND UNAMBIGUOUSLY PROVIDED IN SECTIONS 10.1 AND 10.2, THE SOFTWARE, SERVICE AND CONSULTING SERVICES ARE PROVIDED "AS IS" AND PROVIDER, ITS SUPPLIERS AND SERVICE PROVIDERS DISCLAIM ALL WARRANTIES, EXPRESS OR IMPLIED, INCLUDING, BUT NOT LIMITED TO, IMPLIED WARRANTIES OF MERCHANTABILITY, FITNESS FOR A PARTICULAR PURPOSE, INFORMATIONAL CONTENT, SYSTEM INTEGRATION, ENJOYMENT AND NON-INFRINGEMENT.

Limitation of Liability

Limitation of liability provisions are another tool to control and/or allocate risk. Neither the customer nor the provider wish to be exposed to liability that is potentially unlimited or that cannot be foreseen or quantified. As a result, contracts generally state that liability shall not exceed the value of the contract (or a specified contract or calendar period), or some multiple thereof. While such provisions offer the virtue of simplicity, they may be inadequate for a complex transaction. For example, if a customer pirates the provider's main software product, the losses to the provider could easily exceed the value of the contract. Similarly, if the provider, or an unscrupulous employee of the provider, attempts to re-sell the customer's proprietary information, the costs to the customer could also far exceed the cost of the contract.

Within this context, model Section 11.1 is provider-friendly. It limits the provider's potential liability but does not create a limit for the customer's potential exposure.

Note that in some contracts the limitation of liability section does not apply to any indemnification obligation one party owes the other, or to any remedy set forth elsewhere in the agreement.

Like disclaimer of warranties provisions, exclusion of damages and limitations on liability provisions need to be distinct from the remaining text of the agreement (via use of all caps, boldface type, larger font size, etc.).

11.1 <u>EXCLUSION OF DAMAGES AND LIMITS ON LIABILITY</u>.

NOTWITHSTANDING ANYTHING TO THE CONTRARY IN THE MODEL OR OTHERWISE, PROVIDER, ITS OFFICERS, EMPLOYEES, AFFILIATES, REPRESENTATIVES, CONTRACTORS, SUPPLIERS, PROVIDERS AND SERVICE PROVIDERS WILL NOT BE RESPONSIBLE UNDER ANY CONTRACT OR OTHER THEORY OF RECOVERY (INCLUDING NEGLIGENCE, STRICT LIABILITY OR OTHERWISE) FOR ANY: (A) ERRORS OR INTERRUPTIONS OF USE, LOSSES, INACCURACY OR CORRUPTION OF DATA, OR COST OF PROCUREMENT OF SUBSTITUTE GOODS, SERVICES OR TECHNOLOGY, BUSINESS INTERRUPTIONS OR LOST OPPORTUNITIES; (B) INDIRECT, EXEMPLARY, INCIDENTAL, SPECIAL OR

CONSEQUENTIAL DAMAGES; (C) LOSSES CAUSED BY EVENTS BEYOND PROVIDERS REASONABLE CONTROL (INCLUDING, WITHOUT LIMITATION, THE DISCLOSURE OF CONFIDENTIAL OR OTHER CUSTOMER INFORMATION OR DATA); AND (D) AMOUNTS THAT, IN THE AGGREGATE, EXCEED THE FEES PAID BY CUSTOMER TO PROVIDER FOR THE SERVICES UNDER THE MODEL IN THE 12 MONTHS PRIOR TO THE FIRST ALLEGED ACT OR OMISSION THAT GAVE RISE TO THE LIABILITY.

11.2 <u>EXCLUSIVE REMEDY</u>. CUSTOMER'S EXCLUSIVE REMEDY FOR BREACH OF THE WARRANTY SET FORTH IN SECTION 10.2 WILL BE TO HAVE PROVIDER MAKE A SECOND ATTEMPT TO PERFORM THE CONSULTING SERVICES. IF PROVIDER FAILS TO DO SO AS WARRANTED AFTER NOTICE AND A REASONABLE OPPORTUNITY TO PERFORM, CUSTOMER WILL BE ENTITLED TO RECOVER THE FEES PAID TO PROVIDER FOR THE DEFICIENT CONSULTING SERVICES. THE FOREGOING STATES PROVIDER'S ENTIRE LIABILITY FOR ANY BREACH OF THE WARRANTY SET FORTH IN SECTION 10.2.

Miscellaneous

The "Miscellaneous" section tends to be the contractual equivalent of "flyover country." It is there, but no one really pays attention to it. That is unfortunate, because significant issues may lurk in this section.

Audit

12.1 <u>Audit Rights</u>. Customer shall (i) promptly provide written notice to Provider if the number of users exceeds the maximum number permitted in Exhibit A ("Maximum users"), and (ii) simultaneously pay Provider for any additional users. During normal business hours or at any time the Software or Service is

being used, Provider or its authorized representatives may, upon reasonable advance notice, audit and inspect Customer's use of the Software and Service and/or Customer's compliance with the Agreement.

One such issue is illustrated in model Section 12.1, which permits the provider to audit the customer's use of the services to ensure compliance. Such a provision makes some sense if the contract involves desktop software and the risk of unauthorized copies. Even then, however, the customer would probably insist on more controls on the audit, addressing what records or information the customer is required to produce and what, if anything, the provider may copy. But if the services involve a "pure" ASP or SaaS service (in which the customer merely utilizes the provider's products remotely via the Internet), such an audit provision would seem out of place. Rather, the provider should implement appropriate controls to track the customer's use of the services.

Customer representatives should also recognize that they may need the right to audit the provider, to ensure that agreed security measures are in place or to verify treatment of the customer's confidential or proprietary information. In such event, the customer might insist on provisions similar to these:

A.   *Maintenance of Books and Records*

Provider shall maintain accurate and complete financial records of its activities and operations relating to the Agreement in accordance with generally accepted accounting principles. Provider shall also maintain accurate and complete employment and other records relating to its performance of the Agreement.

B.   *Audits Authorized by Customer*

Provider agrees that Customer, or individuals or entities authorized by Customer (including, without limitation, any regulatory authorities having jurisdiction over Customer or Provider), shall have access to any Provider Service Locations and the right to examine and audit such Provider Service Locations and to examine, audit, excerpt, copy or transcribe any pertinent transaction, activity, or records relating to the Agreement provided such access rights do not constitute an unlawful invasion of the privacy rights of any Provider employee and would not in the reasonable opinion of Provider subject Provider to legal liability. Provider shall provide to such auditors and agents any assistance they may reasonably require in connection with such audits and inspections. All such material, including, but not limited to, all financial records, time cards and other employment records shall be kept and maintained by Provider and shall be made available to Customer during the Term and for a period of five (5) years thereafter unless Customer's written permission is given to dispose of any such material prior to such time.

Assignment

12.2 <u>**Assignment.**</u> **the Agreement will be binding upon and inure to the benefit of the parties to the Agreement and their respective successors and permitted assigns, provided that neither the Agreement nor any license hereunder may be assigned by Customer (whether by operation of law or otherwise) without Provider's prior written consent. Notwithstanding the foregoing, Provider may assign all or any part of its rights and obligations under the Agreement to**

(a) any entity resulting from any merger, consolidation or other reorganization of Provider,
(b) any operating entity controlling Provider, or owned or controlled, directly or indirectly, by Provider,
(c) any affiliate of Provider, or
(d) any purchaser of all or substantially all of the Provider's assets.

As the saying goes, "Change is the only constant." Ownership of companies sometimes changes, or companies change their business focus and attempt to spin off various parts of their corporate structure. Drafters attempt to provide for such possibilities by including language regarding how such changes will affect the contract. Model section 12.2 is an example of such a provision.

Model section 12.2 permits the provider to transfer responsibility for the contract to a successor without any permission from the customer. Any assignment by the customer requires the provider's prior written consent. Occasionally, providers have demanded additional payment as a condition for such consent. In addition, the customer is not protected against transfers to companies that may be its competitors or, what is more likely to occur, to a provider unable to provide the services as required by the agreement. A more equitable allocation of risks with respect to assignment might be:

"The Agreement shall not be assigned by either party without the prior written consent of the other, except that:

(1) Customer may assign the Agreement:

(A) to a parent, a subsidiary corporation, a subsidiary of its parent corporation, or any corporation or entity in which Customer has an ownership interest; or

(B) in the event of an affiliation, merger, acquisition, sale or disposition of substantially all of its assets, consolidation, restructuring, break up, or other joint operating arrangement between Customer and any third party(ies); provided, however, such assignment: is in writing and states that the assignee is accepting the Agreement's material terms, conditions, and obligations as if it were the original party hereto; and

(2) Provider may assign the Agreement to any subsidiary or subsidiary of its parent; provided such assignment:

(A) is in writing; and

(B) states that the assignee is accepting the Agreement's material terms, conditions, and obligations as if it were the original party hereto; provided, however, any such assignee must have the financial solvency and technical expertise to provide the products and/or services required of Provider hereunder to Customer, as determined in Customer's reasonable discretion."

Force Majeure

12.3 <u>Force Majeure</u>. **Neither party may be held liable for any damages or penalty for delay in the performance of its obligations hereunder (other than Customer's obligation to make payments under the Agreement) when the delay is due to the elements, acts of God or other causes beyond its reasonable control.**

Occasionally, performance of a contract is interrupted by events beyond the control of either party, such as storms or disruption of the electrical grid. In such circumstances, it would seem inappropriate to penalize the party that cannot perform, and most contracts contain provisions such as model Section 12.3 that excuse delay caused by forces or events that are beyond the control of the "defaulting party." From the perspective of the customer, however, model Section 12.3 could be unacceptable (particularly if the services in question are critical, as the provider is given no incentive to restore services as quickly as possible).

A provision that gives the customer greater protection for vital processes might read as follows:

> *"Provider recognizes that Customer provides services essential to its business operations, and that these services are of particular importance at the time of a riot, insurrection, civil unrest, natural disaster or similar event. Notwithstanding any other provision of the Agreement, full performance by Provider during any riot, insurrection, civil unrest, natural disaster or similar event is not excused if such performance remains commercially reasonable."*

Arbitration

12.4 <u>Arbitration</u>. **The parties agree that all disputes arising out of or relating to the formation, performance or alleged breach of the Agreement will be determined and settled by binding arbitration to take place exclusively in accordance with the commercial rules of the American Arbitration Association. Any**

award rendered shall be final and binding on the parties, and may be entered as a judgment by any court of competent jurisdiction. The prevailing party shall be entitled to recover its costs of arbitration (including reasonable attorneys' fees), which will be made a part of the arbitrator's award. Notwithstanding the foregoing, in the event irreparable injury is shown, either party may obtain injunctive relief exclusively in the appropriate state or federal court in [insert location of court]. Any litigation between the parties, including litigation to enforce an arbitration award, will take place exclusively in the appropriate state or federal court in [insert location of court].

Unless a contract specifies how disputes will be resolved, the default process is litigation, which tends to be regarded as unnecessarily time-consuming and expensive. It also places the eventual outcome beyond the control of the parties. Consequently, many contracts provide that disputes will be settled by arbitration, not by litigation. Arbitration provisions are enforceable under the laws of most states, and under the Federal Arbitration Act (which specifically directs the federal courts to enforce arbitration provisions found in any contract over which they have jurisdiction). Model Section 12.4 is representative of such provisions and does not favor either the provider or the customer. As a drafting note, one might wish to specify what law (e.g., the laws of which state) will be applied by the arbitration panel. To avoid potential disputes, the parties may also wish to specify where the arbitration is to take place.

More generally, if the contract is large or high value or relates to services vital to the customer, a first step toward resolution of disputes might be to create an escalation mechanism that provides both parties ample opportunity to resolve questions, at the appropriate levels within each party, before the disputes can become potentially disruptive. Representative language might be:

> *"If a dispute arises under the Agreement, then within three (3) business days after a written request by either party, Customer's Project Executive and Providers Project Executive shall promptly confer to resolve the dispute. If these representatives cannot resolve the dispute or either of them determines they are not making progress*

> *toward the resolution of the dispute within three (3) business days after their initial conference, then the dispute may be submitted to the individual designated by Customer and the vice president designated by Provider, who shall promptly confer to resolve the dispute. If the individual designated by Customer and the vice president designated by Provider cannot resolve the dispute, or either one of them determines that they are not making reasonable progress toward resolution of the dispute within five (5) business days after the dispute is first submitted to either the individual designated by Customer or the vice president designated by Provider, then the issue shall proceed pursuant to the Formal Resolution process described in Section _____."*

Severability

12.5 Severability. A determination that any provision of the Agreement is invalid or unenforceable will not affect the validity or enforceability of any other part of the Agreement. Similarly, a determination that any provision is invalid or unenforceable in one application will not affect the validity or enforceability of the same provision in other contexts. To the extent possible, the Agreement shall be construed to give meaning to every provision.

If the contract is eventually the subject of litigation or arbitration, it is typically in both parties' best interests to ensure that, if any specific clause is found not to be enforceable or valid, such unenforceability or invalidity will not affect the other provisions of the agreement. Most states also allow reformation of specific clauses to effectuate the expressed intent of the parties in the absence of such clause.

Waiver and Modification

12.6 Waiver and Modification. A party's waiver of any breach or its failure to enforce any term of the Agreement may not be deemed a waiver of any other breach or of its right to enforce the same term or others in the future. Any waiver, amendment, supplementation or other modification or

> **supplementation of any provision of the Agreement will be effective only if in writing and signed by both parties.**

For the avoidance of doubt, model Section 12.6 expressly provides that one waiver will not affect future rights for non-performance and that all remedies remain available to both parties. This section also requires that any waiver or amendment be in writing and signed by both parties (as opposed to simply the "party to be charged" as allowed in some jurisdictions).

Governing Law

> **12.7 <u>Governing Law</u>. the Agreement will be governed by and construed in accordance with the substantive laws of the United States and the State of [insert state], without regard to or application of [insert state]'s conflicts of law rules. the Agreement will not be governed by the United Nations Convention on the International Sale of Goods or the Uniform Computer Information Transactions Act, the application of which are expressly excluded.**

Assume the headquarters of the provider are in New York, and those of the customer are in California. Their representatives meet in Chicago and sign an agreement for services to be performed in Texas. Is the contract governed by the laws of New York, California, Chicago, or Texas? To resolve such questions, commercial contracts typically specify what law will apply. Naturally, each party prefers the law of its state, but most sophisticated parties will be willing to accept the law of a "neutral" or "mutually inconvenient" territory such as New York. Note that model Section 12.7 excludes application of the Uniform Computer Information Transaction Act. As such, the provision is relatively customer-friendly, as the act is widely regarded as a "provider protection act."

While model Section 12.7 is silent, note that choice of laws may need to be paired with selection of the appropriate venue and jurisdiction (which determine *where* the contract will be adjudicated, as opposed to which laws will be used to enforce the contract).

Notices

12.8 <u>Notices</u>. All notices required or permitted under the Agreement must be in writing, must reference the Agreement and will be deemed given: (i) when sent by facsimile with a confirmation page generated by the sending device; (ii) 5 business days after having been sent by registered or certified mail, return receipt requested, postage prepaid; or (iii) I working day after deposit with a commercial overnight carrier, with written verification of receipt. To be effective, a confirmation copy of a notice must be sent contemporaneously via U.S. mail. All communications must be sent to the contact information set forth below or to such other contact information as may be designated by a party by giving written notice to the other party pursuant to this section 12.8:

To Provider: **To Customer:**

_____   _____

Attn: Legal Notices   **Attn: Legal Notices**

Phone: _____   **Phone:** _____

Fax: _____   **Fax:** _____

A notice provision is particularly valuable if the contract involves large parties with many offices or divisions. If a notice requires immediate action, sending it to the main post office box of a large multinational corporation might not be the most efficient course. A careful drafter will therefore take the time to identify those addresses and individual contacts within each party responsible for managing performance of the contract (as well as the mode of remittance to avoid inadvertent missteps with respect to notice).

Relationship of Parties

12.9   Relationship of Parties. The Agreement will not be construed as creating an agency, partnership, joint venture or any other form of legal association between the parties and each party is an independent contractor.

Occasionally, contracts between parties give rise to the impression that one party is the agent of the other, or that the personnel of one party are employees or representatives of the other. An agent, of course, has legal authority to bind its principal, while an employer (rather than a principal) incurs certain legal liabilities, such as payroll taxes and workers compensation insurance. To preclude such questions, drafters routinely include provisions such as model Section 12.9. Note, however, that sometimes a customer will want a provider to act as a limited agent. For instance, a customer may desire that the provider secure certain products on the customer's behalf. If such an agency appointment is to occur, the drafter should take care to specifically define the limits of the provider's authority to bind the customer.

Attorneys' Fees

12.10 Attorneys' Fees. In any action or proceeding to enforce rights under the Agreement, the prevailing party will be entitled to recover its costs and attorneys' fees.

Under the "American rule" of litigation, each party is expected to pay its own expenses, including the cost of its own attorneys, unless the parties agree otherwise. Model Section 12.10 is an example of such an agreement. Typically, such provisions are regarded as a safeguard against legal actions that have no merit (in that a party should be less likely to institute a legal proceeding if they believe they could lose *and* be required to pick up the tab for the other side's legal team).

Construction

12.11 Construction. The Agreement shall be deemed the joint work product of the parties and may not be construed against

either party as drafter. Captions are for convenience only and may not be construed to define, limit, or affect the construction or interpretation of the Agreement.

One of the rules of contract law is that, in the event a contract is ambiguous, the ambiguity will be construed against the drafter (who presumably was in the best position to make the contract clear and precise). To avoid this rule, commercial contracts typically contain a construction clause that explicitly states that the agreement is a joint work product, not to be construed against either party. Further, the construction provision ensures that captions will not negate the actual text incorporated under them.

Entire Agreement

12.12 Entire Agreement. The Agreement, including Exhibits and any order form or Statement of Work, constitutes the entire agreement between the parties with respect to the subject matter of the Agreement, and supersedes and replaces all prior or contemporaneous written or oral statements, understandings or agreements. Except where otherwise set forth in the main body of the Agreement, in the event of a conflict between an Exhibit or Statement of Work and the main body of the Agreement, the terms of the main body of the Agreement will prevail.

Model section 12.12 is a typical "integration clause," providing that the entire agreement between the parties is set forth in this particular written contract, its exhibits, and any subsequent SOWs. Integration clauses are intended to preclude claims that the written agreement does not capture the entire agreement and that it does not express the intent or the actual agreement of the parties. Such claims are usually supported by documents from the negotiation file or testimony from members of the negotiation team. In effect, such claims invite the court (or arbitrator) to reopen the contract negotiations (*ex parte*). Integration clauses foreclose such reliance on documentation extrinsic to the text of the agreement by stating affirmatively that the contract does indeed reflect the final agreement of the parties, and other materials are irrelevant and should be ignored.

The Signature Block

Provider	**CUSTOMER**
By _____	**By** _____
Name _____	**Name** _____
Title _____	**Title** _____
Date _____	**Date** _____

Who may sign the contract is an important, but often over-looked, question. If the person who signs does not have appropriate authority, the contract may not be enforceable (although the person who signs it may be personally liable). For this reason companies may require that the contract be executed by an officer of the other party. On large deals, the parties may require certificates of incumbency—formal evidence that the person assigned to sign the contract is authorized to do so.

Exhibit A
Services and Support Ordered by Customer

As mentioned above, a "master agreement" can be divided into roughly two parts:

- The "master" itself, which sets out basic terms and conditions;
- The exhibits (or "attachments" or "schedules"), which set out the details of specific transactions or orders between customer and provider.

Exhibits are typically subject to the terms of the master, unless specific exceptions or changes are set out in a particular exhibit.

The master agreement may describe generally the goods or services available from provider, and may set out pricing, or price controls. The exhibits will spell out exactly what customer wants, and when, and what customer will pay for it.

Astute customers will insist on language stating that they are not obligated to purchase any goods or services under the master; that they do not incur any financial obligation until a mutual agreed exhibit is executed.

1.   [Name of Services] Ordered by Customer

Service	Pricing	Quantity	Total Fees

2.   Initiation Fee

Upon execution of this Agreement, Customer shall incur an initiation fee of $____, due and payable no later than 30 days after the Effective Date.

Initiation fees are frequently encountered, but an astute customer might well ask, "What am I paying for? Isn't setting up a new customer and throwing the necessary switches part of the provider's overhead?"

3.   Support Fee

Initial Support Level Purchased (Check One):

Standard Support

Gold Support

Support Fee:

> **$_____ for service through dd/mm/yyyy. Thereafter, support fees will be at Provider's current rates. Provider's support fees as of the Effective Date for the Services purchased by Customer are $_____/ per calendar quarter, but are subject to change upon 30 days notice from Provider. Customer shall pay all support fees in advance, no less than 30 days prior to the first day of the calendar quarter for which the support services are purchased.**

A customer might have several reservations about this language:

- What are "Provider's current rates"? A better course for the customer might be to include the applicable rates or rate card, so the customer is able to budget appropriately.

- This language would permit the provider to increase fees every month, making it difficult for the customer to budget, let alone control its costs. Better language, from the perspective of a customer, would be to permit increases less frequently—perhaps once a year—and to limit the size of the increase (e.g., "no more than X percent per annum").

Change Fee for Downgrade in Support Level:

In the event Customer wishes to reduce or downgrade its service level, Customer shall incur a one-time fee of $____.

As mentioned earlier, paying to reduce the level of support is counterintuitive. Most customers could be expected to question such an expense.

4.   Locations

Customer shall be authorized to access the Services from the following locations::_____.

As discussed above, customer may be unduly hampered if able to access provider's services only from certain locations. On the other hand, if provider is to provide services or support on-site, it is appropriate to specify where those services will be provided.

Exhibit B
Support Terms and Conditions

Provider offers different levels of support services, for which it charges different fees. Provider reserves the right to change the level or quality of support upon thirty (30) days written notice.

Permitting the provider to change support on thirty days written notice significantly reduce the value the contract has to the customer—who no longer has a meaningful commitment regarding the level, quality, or cost of the services to be provided. Customers may well require that support change no more frequently than annually, or that no change will reduce the quantity or quality of support provided or increase the price the customer has previously agreed to.

1. Support Provided

   a. <u>Support Hours</u>. Provider will provide telephone consultation and advice to Customer and respond to e-mail and fax messages sent to Provider's support department: (i) for standard support, between the hours of [insert hour] a.m. and [insert hour] p.m., Pacific Time, Monday through Friday, excluding holidays; and (ii) for gold support, between the hours of [insert hour] a.m. and [insert hour] p.m., Central Time, Monday through Saturday, excluding holidays.

   b. <u>Limitations</u>. Customer acknowledges that Provider is under no obligation to support any hardware or software that is not part of the Service. Response times are based upon the severity level of Customer's problem, as determined by Provider and described in Section 2 of this Exhibit B.

Note that support is apparently not available twenty-four hours a day, seven days a week. If the customer's operations span many time zones, the provider's preferred hours may not be acceptable.

2. Service Guidelines.

a. **Problem Classification.** Provider will assign a severity level to each problem reported by Customer, based on the Problem Classification Table below. Severity 3 will be the default severity level, unless otherwise specified by Provider's support personnel. Severity levels are defined as follows:

Not all problems are created equal; neither do they all require immediate correction. For that reason services contracts typically divide problems into different categories, ranging from "critical" (or similar term) to "minor," and set out different response times and procedures. This approach permits providers to make the most efficient use of its resources and personnel, and keeps costs down for customer. Care should be taken, however, to ensure that the specified response times meet the needs of customer. Customer should also recognize that, the more resources it requires from provider, the greater the cost will be.

Problem Classification Table

Problem Classification	Criteria
Severity 1 (Critical)	The Service is completely unavailable or unusable. The problem affects time-critical applications, without which Customer cannot conduct business. No known work-around is currently available.
Severity 2 (Serious)	The Service is significantly impaired. Customer cannot conduct its integral business. No known work-around is currently available.
Severity 3 (Degraded)	The Service is not functioning in accordance with its specifications. Integral business processes of Customer, however, have not been interrupted.
Severity 4 (Minimal)	The Service problems have little or no impact on Customer's daily business process.

b. <u>Error Reporting and Response</u>. For all (i) e-mail and fax support and (ii) Customer problems which cannot be resolved in an initial telephone support conversation, Customer must provide Provider with a reasonably detailed description of the problem by e-mail or fax. Provider will take the following steps, in accordance with the Response Expectation Tables below:

<u>Step 1</u>   Provider will acknowledge Customer's problem and begin collecting additional information from Customer.

<u>Step 2</u>   Provider will actively address the problem during support hours (as specified above) and provide a temporary patch, correction, or workaround as soon as reasonably possible.

<u>Step 3</u>   Provider will provide a permanent solution in the form of a tested permanent patch or a completely new release of the applicable software.

Customer must provide Provider with contact information for employees who will be available outside of Customer's normal business hours if any support will be provided at such times.

Provider will use reasonable efforts to respond to Customer's support problems during the response times indicated in the Response Expectation Table. In most cases, Provider will respond as follows:

Response Expectation Table (Standard Support)

Severity	Step 1	Step 2	Step 3
1 (Critical)	4 support hours	Immediate and continuing effort during support hours	Within 60 calendar days
2 (Serious)	8 support hours	1 to 5 business days	Within 90 calendar days

| 3 (Degraded) | 16 support hours | Within 10 business days | Next scheduled release of software |
| 4 (Minimal) | 24 support hours | Worked on a time available basis | If Provider determines this step to be necessary |

Response Expectation Table (Gold Support)

Severity	Step 1	Step 2	Step 3
1 (Critical)	2 support hours	Immediate and continuing effort 24 hours a day	Within 30 calendar days
2 (Serious)	4 support hours	2 business days	Within 45 calendar days
3 (Degraded)	8 support hours	Within 5 business days	Next scheduled release of software
4 (Minimal)	16 support hours	Worked on a time available basis	If Provider determines this step to be necessary

c. <u>Escalation Process</u>. If the Provider support representative on duty at the time Customer reports a problem cannot correct any level 1 to 3 problem or implement a plan of resolution within the time set forth in the Escalation Table below, (s)he will notify support management, which will do so. At each stage in the Escalation Table below, Customer shall make available to Provider a Customer contact at an equivalent management level, who has the authority to make decisions about alternative approaches for resolving Customer's problem. In addition, for level 1 problems, Customer shall make available to Provider a

Customer contact who will be continuously available to assist Provider's support personnel with data gathering, testing, and applying fixes. Upon Provider's request, Customer shall provide access to its computing environment if Provider cannot duplicate Customer's problem in-house.

Provider will make reasonable efforts to notify its support management, according to the following schedule:

Escalation Table

Elapsed Time	Severity 1 (Critical)	Severity 2 (Serious)	Severity 3 (Degraded Operations)
Immediately	Support Group Leader		
2 hours	Support Manager		
4 hours	Support Director	Support Group Leader	
8 hours			
16 hours	VP of Development	Support Manager	
24 hours			Support Group Leader
48 hours		Support Director	
72 hours	CEO	VP of Development	Support Manager

Elapsed Time represents the number of support hours (not clock hours) that have passed since the problem was reported by Customer.

Exhibit C
Consulting Services Order Form

This Exhibit is an example of "just the facts" drafting. It lays out what will be provided, where, when, and at what price. It is customer-friendly in that it gives customer broad rights to reject provider personnel and the right to terminate the order, or portions of it, with or without cause, but without a penalty. Most vendors will resist that provision.

The Exhibit also contains at least one significant omission—it does not address ownership of work product. If provider creates a new process that gives customer a major competitive advantage, will the intellectual property rights belong to customer or to provider?

This Order is entered into as of the __ day of _____, 20__, between _____("Customer") and _____ ("Provider").

Whereas, Customer and Provider executed a Master Agreement having an Effective Date of _____ ("Agreement"), for the acquisition of computer products and/or related services from Provider by Customer;

Now, therefore, in consideration of the mutual covenants, agreements, and conditions contained herein, Customer and Provider agree as follows:

1.   PRODUCTS/SERVICES:

QTY.	TYPE	PRICE/ LICENSE FEE	ANNUAL MAINT/ SUPPORT CHARGE	DELIVERY /SERVICES DATE	INSTALL DATE

2. **LOCATION:**

The equipment/software/services shall be delivered/installed/performed at the following location:

Ship to:	Bill to:
Attn: _____	Attn: _____

3. **TECHNICAL PERSONNEL:** All technical personnel contracted to Customer by Provider that are deemed by Customer to be unsatisfactory for any reason will be immediately removed by Provider. Provider shall notify Customer, within 24 hours, of either a removal or discontinuance, whether and when a suitable replacement will be provided, or whether Provider will be unable to provide a suitable replacement. Customer, in its sole discretion, shall have the right to accept or reject any replacement personnel offered by Provider. Customer shall provide on-site technical personnel with adequate work space, and necessary clerical services and incidental supplies at no cost to Provider.

4. **ORDER TERMINATION BY CUSTOMER:** Customer may terminate any Services pursuant to this Order, with or without cause, by giving 30 days prior written notice to Provider.

CUSTOMER   **(Provider)**

BY: _____   BY: _____

NAME: _____   NAME: _____

TITLE: _____   TITLE: _____

DATE: _____   DATE: _____

Appendix A:
Complete Master
Subscription Agreement

This Master Subscription Agreement ("Agreement") is made and entered into as of [insert effective date] (the "Effective Date"), by and between [insert Name of Provider] ("Provider"), a [insert state of incorporation] Corporation, with offices at [insert location of offices], and [insert Name of Licensee] ("Licensee") a [insert state of incorporation] Corporation, with offices at [insert location of offices] ("Customer").

1. DEFINITIONS

1.1   "Consulting Services" will have the meaning set forth in section 3.1.

1.2   "Confidential Information" will have the meaning set forth in sections 5.1 and 5.2.

1.3   "Customer Data" means all information provided by Customer to Provider through the Service for use in conjunction with the Services and the Software, including processing, storage and transmission as part of the Services.

1.4   "Customer Information" means all information created or otherwise owned by Customer or licensed by Customer from third parties, including Customer Data and information created by Customer by using the Services, that is used in conjunction with the Services and the Software.

1.5   "Documentation" means all configurations and specifications published by Provider from time to time relating to the Software or the Services.

1 .6   "Equipment" will have the meaning set forth in section 4.2.

1.7   "Locations" means the physical location or locations set forth in Exhibit A from which Customer is licensed to access the Service.

1.8   "Maximum Users" will have the meaning set forth in section 12.1.

1 .9   "Provider Information" means all information, including the Software, created or otherwise owned by Provider or licensed by Provider from third parties, related to the Services and any materials prepared by Provider pursuant to a Statement of Work under this Agreement.

1.10   "Services" means Provider's electronic data processing, storage and transmission services ordered by Customer, which are enumerated in Exhibit A.

1.11   "Software" means the software used by Provider to provide the Services.

1.12   "Statement of Work" will have the meaning set forth in section 3.1.

2.   SERVICES AND SUPPORT

2.1   Provision of Services. Subject to the terms and conditions of this Agreement, Provider will use reasonable commercial efforts to provide the Services to Customer. Customer may request additional services from Provider, but only by submitting a written request to Provider. If Provider accepts Customer's request, then Provider shall provide the additional Services on the terms set forth in this Agreement. Provider will have no obligation to provide any upgrades to the Software. During the term of this Agreement, Provider may make enhancements to the Software and the Services and Customer agrees to use the enhanced versions of the Software and the Services.

2.2   Grant of License. Subject to the terms and conditions of this Agreement, Provider grants to Customer a limited, non-transferable, non-exclusive license for the term of this Agreement to access via the Internet and use the Services and the Software, but only from the Locations and solely to support Customer's normal course of business.

2.3   Restrictions on Use. Customer may not, directly or indirectly, (i) license, sublicense, sell, resell, lease, assign, transfer, distribute, or otherwise commercially exploit or make available to any third parties the Services or the Software in any way, (ii) alter, modify, translate or create derivative works based on the Services or the Software, (iii) process or permit to be processed the data of any third party, (iv) use or permit the use of the Services or the Software in the operation of a service bureau, timesharing arrangement or otherwise for the benefit of a third party, (v) disassemble, decompile, or reverse engineer the Software or any aspect of the Services, or otherwise attempt to derive or construct source code or other trade secrets from the Software, (vi) build a competitive product or service to the Services or Software or a product or service that uses similar ideas, features, functions, or graphics to the Services or Software, (vii) use the Services or Software to engage in any prohibited or unlawful activity, or (viii) permit any third party to do any of the foregoing.

2.4   Support Services. Subject to Customer's prompt payment of the fees due under this Agreement, Provider shall provide Customer with the support services specified in Exhibit A. Provider's current support services are set forth in Exhibit B. Customer may upgrade to a higher level of support services without paying a change fee. However, Customer must pay the change fee set forth in Exhibit A to downgrade or cancel support services.

2.5   Use of Customer Data/Customer Representations and Warranties. Customer shall be solely responsible for collecting, inputting and updating all Customer Data. Customer represents and warrants that its Customer Information does not and will not include anything that infringes the copyright, patent, trade secret, trademark or any other intellectual property right of any third party; contains anything that is obscene, defamatory, harassing, offensive, malicious or which constitutes child pornography; or otherwise violates any other right of any third party.

2.6   Passwords. Provider shall provide Customer with passwords to access the Service. Customer shall be responsible for all use of its account(s). Customer shall also maintain the confidentiality of all passwords assigned to it. Customer may not share its passwords with third parties or attempt to access the Service without providing a password assigned to it.

Passwords cannot be shared or used by more than one individual, but may be reassigned from time to time to new individuals who are replacing former individuals who have terminated employment with Customer or otherwise changed job status or function such they no longer have a need to use the Service or Software. In no event shall Customer allow more individuals to access the Software or Services than the maximum number of passwords provided to Customer under this Agreement and Exhibit A.

3.   CONSULTING SERVICES

3.1   Statements of Work. Upon Customer's request and subject to both parties' acceptance of a Statement of Work, Provider may modify, customize or enhance the Software or provide implementation, data transfer, training, or other services relating to the Software or Services ("Consulting Services"), for an additional fee. Customer shall request Consulting Services by completing a Provider order form, in the form attached as Exhibit C. Upon receipt of Customer's written request for Consulting Services, Provider shall prepare a Statement of Work. A Statement of Work is an offer to perform Consulting Services on specified terms and for specified fees. A Statement of Work will only he binding if signed by both parties. Each Statement of Work will be governed by this Agreement. In the event of a conflict between a Statement of Work and this Agreement, the terms and conditions of this Agreement will prevail.

3.2   Payment for Services. Unless a Statement of Work expressly states that Consulting Services are to be performed for a fixed fee, Provider shall provide Consulting Services to Customer on a "time and materials basis" with any price quotation constituting merely an estimate that is not a binding on Provider. For purposes of this Agreement, "time and materials basis" means that Customer shall pay Provider at the hourly rates specified by Provider for the Consulting Services and shall reimburse Provider for expenses as set forth in section 3.3.

3.3   Expenses. In addition to payment for the Consulting Services, Customer shall reimburse Provider for reasonable travel, administrative, equipment, and out-of-pocket expenses incurred in performing the Services, in accordance with section 6. These expenses are not included in any estimate in a Statement of Work unless expressly itemized.

3.4   Effect of Termination during Performance of Consulting Services. Upon termination of this Agreement, Provider will be relieved of all obligations to perform Consulting Services, the Statement(s) of Work for all Consulting Services will terminate, and Customer immediately shall pay Provider for Consulting Services performed prior to the date of termination. Should Customer terminate any Statement of Work prior to the conclusion of the Statement of Work, Customer shall reimburse Provider for any costs associated with winding down Consulting Services under that terminated Statement of Work.

3.5   Termination of Statement of Work/Right to Cure.

(i)   Either party may terminate the Statement of Work if the other party breaches a Statement of Work and has not cured its breach within 30 days of written notice from the non-breaching party. In addition, Provider may terminate the entire Agreement if Customer breaches any payment obligation. Otherwise, however, neither party may terminate the entire Agreement based solely on breach of a Statement of Work.

(ii)   Consent to extend the cure period for breaches of a Statement of Work may not be unreasonably withheld so long as the breaching party has in good faith begun to cure its defects in performance during the 30 day notice period. Notwithstanding the foregoing, Provider, in its sole discretion, may withhold its consent to extend the cure period for Customer's breach of its payment obligations.

3.6   Change Orders. Provider's obligation to perform Consulting Services is limited to the services described in a Statement of Work. Customer may request changes to a Statement of Work by submitting proposed changes in writing to Provider. Within 14 days of receiving Customer's request, Provider will provide Customer with a proposed change order. A change order will set forth, at a minimum, (i) a written description of the changes to the Statement of Work, (ii) any changes to the services schedule, and (iii) any changes to the fee estimate. A change order will be binding only if signed by both parties, in which case it will be governed by the terms and conditions of this Agreement. In the event of a conflict between a change order and a Statement of Work, the Statement of Work will prevail.

3.7   Cooperation. Customer shall cooperate with and assist Provider in performing Consulting Services. Among other things, Customer shall provide to Provider timely access to office accommodations, facilities, and equipment; complete and accurate information and data from its officers, agents, and employees; and suitably configured computer products. Customer's failure to do so will relieve Provider of responsibility for any related deficiencies in its performance.

4.   CUSTOMER RESTRICTIONS AND RESPONSIBILITIES

4.1   Compliance with Documentation and Laws/Representations and Warranties. Customer represents, covenants, and warrants that Customer will use the Services and any data of third parties only as contemplated by the Documentation and in compliance with all applicable laws and policies (including hut not limited to laws, government regulations, Provider policies and any other applicable policies relating to intellectual property, employment, labor, spamming, spoofing, network security, privacy, obscenity or defamation).

4.2   Customer Equipment. Customer shall be responsible for selecting, obtaining and maintaining any equipment and ancillary services needed to connect to, access or otherwise use the Services, including, without limitation, modems, hardware, servers, software, operating systems, networking equipment, web servers, and long distance and local telephone service, but excluding the Software (collectively "Equipment"). Customer shall be responsible for ensuring that the Equipment is compatible with the Services and the Software and complies with the Documentation. Customer shall also be responsible for the security and use of the Equipment.

4.3   Customer Requirements. Customer shall be responsible for determining what Services, Software and Consulting Services and support it requires.

5.   CONFIDENTIALITY AND PUBLICITY

5.1   Confidentiality Obligation. "Confidential Information" means all written or oral information designated as confidential at the time of disclosure that is made accessible to the other party in connection with this

Agreement including, without limitation, computer programs, software, formulas, data, information, inventions, techniques, strategies, trade secrets, know-how, plans for products or services, marketing plans, financial documents or data, processes and designs, Service passwords, and the terms, hut not the existence of, this Agreement. Written Confidential Information must be marked as "confidential" or "proprietary." Oral Confidential Information must be designated as confidential at the time of disclosure and reduced to a written summary and marked "confidential" or "proprietary" within 10 days of the oral disclosure. Each of the parties shall treat the other party's Confidential Information confidentially and with at least the same degree of care it uses to prevent the disclosure of its own Confidential Information, but in no event less than reasonable care. In addition, each party shall use the Confidential Information of the other party solely in the performance of its obligations under this Agreement and not disclose it, except to authorized employees of the receiving party or its affiliates, its legal counsel and its accountants (provided that the receiving party contractually obligates them to a duty of confidentiality no less restrictive that the duty imposed by this section 5.1 and remains jointly and severally liable for any breach of confidentiality by them). Each party shall promptly notify the other party of any actual or suspected misuse or unauthorized disclosure of its Confidential Information. Upon expiration or termination of this Agreement, each party shall return all tangible copies of any Confidential Information received from the other party. To the extent that relevant federal or state laws or regulations may specify a higher level of confidentiality than set out in this Agreement with respect to certain forms of Confidential Information, such higher standard shall govern as between the parties.

5.2   Exclusions. Confidential Information will not include information that the recipient can prove: (i) was generally available to the public at the time it was disclosed, (ii) was known to the recipient, without restriction, at the time of disclosure by the disclosing party, (iii) is disclosed with the prior written approval of the disclosing party, (iv) was independently obtained or developed by the recipient without any use of the Confidential Information, (v) becomes known to the recipient, without restriction, from a source other than the disclosing party who does not owe a duty of confidentiality to the disclosing party and obtained the information by lawful means, or (vi) is disclosed in response to an order or requirement of a court, administrative agency, or other governmental body, a subpoena, or by the rules of a securities market or exchange on which the disclosing party's securities are traded (but only if (a) the recipient vigorously opposes the proposed disclosure and provides prompt advance notice to the disclosing party to enable it to appear and independently contest the disclosure, and (b) any Confidential Information so disclosed will otherwise remain subject to the provisions of this section 5). The burden of proof in establishing that any Confidential Information is subject to any of the foregoing exceptions will be borne by the receiving party.

5.3   Trademark License. Customer grants Provider a limited, nonexclusive, worldwide right and license during the Term to use the names, logos and trademarks of Customer to publicize the existence of the business relationship established by this Agreement.

6.   PAYMENT OF FEES

6.1   Fees. Customer shall pay Provider the fees for the Services as set forth on Exhibit A. Provider reserves the right to change the fees or applicable charges upon 30 days prior notice by e-mail or otherwise. Unless otherwise specified in Exhibit A, Customer shall pay the fees set forth in the invoice to Provider within 30 days of the invoice date.

6.2   Disputed Amounts. Provider need not provide any adjustments or credits to Customer for disputed amounts billed by Provider unless Customer provides written notice of the disputed amounts to Provider within 60 days after the invoice date on the first invoice in which the error or problem appeared. Customer must direct any notices required under this section 6.2 (and any related inquiries) to Provider's customer support department.

6.3   Interest and Taxes. Unpaid fees are subject to a finance charge of [insert interest rate] percent per month on any outstanding balance, or the maximum permitted by law, whichever is lower, plus all expenses of collection (including reasonable attorneys' fees). Customer shall he responsible for taxes associated with the Services or Consulting Services other than for appropriate sales and use taxes.

7.   TERM AND TERMINATION

7.1   <u>Term</u>. Unless terminated earlier in accordance with section 7.2, the term of this Agreement will be [insert term] years. This Agreement may be renewed by mutual written agreement of the parties.

7.2   <u>Termination</u>. This Agreement may be terminated as follows:

(i)   Customer may terminate this Agreement at any time by notifying Provider in writing and paying all undisputed fees for the Services for the remainder of the then-current term.

(ii)   Either party may terminate this Agreement immediately by written notice if the other party materially breaches this Agreement and fails to cure its breach after receipt of written notice within (a) 15 days in the case of non-payment of any fees, or (b) 30 days in the case of all other breaches.

(iii)   Either party may terminate this Agreement immediately by written notice if the other party (a) becomes insolvent, (b) makes an assignment for the benefit of creditors, (c) files or has filed against it a petition in bankruptcy or seeking reorganization, (d) has a receiver appointed, or (e) institutes a proceeding for liquidation or winding up. In the case of involuntary proceedings, a party will only be in breach if the applicable petition or proceeding has not been dismissed within 90 days.

7.3   <u>Effect of Termination</u>. Within 7 days of expiration or earlier termination of this Agreement, Customer shall pay to Provider all undisputed fees for the Services up to and including the date of termination.

7.4   <u>Survival</u>. The provisions of sections 2.3, 2.5, 2.6, 3.2, 3.3, 3.4, 4.1, 5.1, 5.2 and 6 though 12 will survive the expiration or earlier termination of this Agreement. Customer's obligations under section 5 with respect to the Software and Services shall survive the expiration or earlier termination of this Agreement for a period of 2 years.

8.   PROPRIETARY RIGHTS

8.1   Provider's Proprietary Rights. Exclusive of Customer Information, Provider (or it's third-party licensors, if applicable) will retain all rights, title, and interest in and to the Software, Services, and the Provider Information and all legally protectable elements or derivative works thereof. Provider may place copyright and/or proprietary notices, including hypertext links, within the Services. Customer may not alter or remove these notices without Provider's written permission. Customer may not have the right to, and agrees not to, attempt to restrain Provider from using any skills or knowledge of a general nature acquired during the course of providing the Services, including information publicly known or available or that could reasonably be acquired in similar work performed for another Provider customer.

8.2   Customer's Proprietary Rights. Customer will retain all rights, title and interest in and to the legally protectable elements of Customer Information and derivative works thereof.

9.   INDEMNITY

Customer will indemnify, defend, and hold Provider harmless from and against any and all costs, liabilities, losses and expenses (including, but not limited to, reasonable attorneys' fees) resulting from any claim, suit, action or proceeding brought by any third party against Provider arising out of or relating to Customer's breach of its representations or warranties hereunder or its use of the Software and Services.

10.   WARRANTY AND DISCLAIMER

10.1   Limited Warranty for Services. Provider shall use reasonable commercial efforts consistent with prevailing industry standards to maintain the security of the Services and minimize errors and interruptions in the Services, provided that:

(i)   Customer uses the Service and the Software strictly in accordance with the Documentation,

(ii)   Customer pays all amounts due under this Agreement and is not in default of any provision of this Agreement, and

(iii)   Customer makes no changes (nor permits any changes to be made other than by or with the express approval of Provider) to the Software or Service.

In addition, Customer acknowledges that the Services may be temporarily unavailable for scheduled maintenance, for unscheduled emergency maintenance, or because of other causes beyond Provider's reasonable control (including without limitation delays or other problems inherent in the user of the Internet and electronic communications services). Provider will not be liable to Customer as a result of these temporary service interruptions.

10.2   **Limited Warranty for Consulting Services.** Provider warrants to Customer that any Consulting Services will be performed in a manner consistent with generally accepted industry practices. In order to receive this warranty, Customer must report any deficiencies in the Consulting Services to Provider in writing within 90 days of completion of the services identified in a particular Statement of Work. Any deficiencies not reported within such 90 day period shall be deemed waived by Customer.

10.3   DISCLAIMER OF WARRANTIES FOR SERVICE AND SOFTWARE. NEITHER PROVIDER NOR ITS SUPPLIERS OR SERVICE PROVIDERS WARRANT THAT THE SERVICES WILL BE UNINTERRUPTED OR ERROR FREE, NOR DO THEY MAKE ANY WARRANTY ABOUT THE RESULTS THAT MAY BE OBTAINED BY USING THE SOFTWARE OR SERVICES. EXCEPT AS EXPRESSLY AND UNAMBIGUOUSLY PROVIDED IN SECTIONS 10.1 AND 10.2, THE SOFTWARE, SERVICE AND CONSULTING SERVICES ARE PROVIDED "AS IS" AND PROVIDER, ITS SUPPLIERS AND SERVICE PROVIDERS DISCLAIM ALL WARRANTIES, EXPRESS OR IMPLIED, INCLUDING, BUT NOT LIMITED TO, IMPLIED WARRANTIES OF MERCHANTABILITY, FITNESS FOR A PARTICULAR PURPOSE, INFORMATIONAL CONTENT, SYSTEM INTEGRATION, ENJOYMENT AND NON-INFRINGEMENT.

11.   LIMITATION OF LIABILITY

11.1   EXCLUSION OF DAMAGES AND LIMITS ON LIABILITY. NOTWITHSTANDING ANYTHING TO THE CONTRARY IN THIS AGREEMENT OR OTHERWISE, PROVIDER, ITS OFFICERS, EMPLOYEES, AFFILIATES, REPRESENTATIVES, CONTRACTORS, SUPPLIERS, PROVIDERS AND SERVICE PROVIDERS WILL NOT BE RESPONSIBLE UNDER ANY CONTRACT OR OTHER THEORY OF RECOVERY (INCLUDING NEGLIGENCE, STRICT LIABILITY OR OTHERWISE) FOR ANY: (A) ERRORS OR INTERRUPTIONS OF USE, LOSSES, INACCURACY OR CORRUPTION OF DATA, OR COST OF PROCUREMENT OF SUBSTITUTE GOODS, SERVICES OR TECHNOLOGY, BUSINESS INTERRUPTIONS OR LOST OPPORTUNITIES; (B) INDIRECT, EXEMPLARY, INCIDENTAL, SPECIAL OR CONSEQUENTIAL DAMAGES; (C) LOSSES CAUSED BY EVENTS BEYOND PROVIDER'S REASONABLE CONTROL (INCLUDING, WITHOUT LIMITATION, THE DISCLOSURE OF CONFIDENTIAL OR OTHER CUSTOMER INFORMATION OR DATA); AND (D) AMOUNTS THAT, IN THE AGGREGATE, EXCEED THE FEES PAID BY CUSTOMER TO PROVIDER FOR THE SERVICES UNDER THIS AGREEMENT IN THE 12 MONTHS PRIOR TO THE FIRST ALLEGED ACT OR OMISSION THAT GAVE RISE TO THE LIABILITY.

11.2   EXCLUSIVE REMEDY. CUSTOMER'S EXCLUSIVE REMEDY FOR BREACH OF THE WARRANTY SET FORTH IN SECTION 10.2 WILL BE TO HAVE PROVIDER MAKE A SECOND ATTEMPT TO PERFORM THE CONSULTING SERVICES. IF PROVIDER FAILS TO DO SO AS WARRANTED AFTER NOTICE AND A REASONABLE OPPORTUNITY TO PERFORM, CUSTOMER WILL BE ENTITLED TO RECOVER THE FEES PAID TO PROVIDER FOR THE DEFICIENT CONSULTING SERVICES. THE FOREGOING STATES PROVIDER'S ENTIRE LIABILITY FOR ANY BREACH OF THE WARRANTY SET FORTH IN SECTION 10.2.

12.   MISCELLANEOUS

12.1   Audit Rights. Customer shall (i) promptly provide written notice to Provider if the number of users exceeds the maximum number permitted in Exhibit A ("Maximum users"), and (ii) simultaneously pay Provider for any additional users. During normal business hours or at any time the Software or Service is being used, Provider or its authorized representatives may, upon reasonable advance notice, audit and inspect Customer's use of the Software and Service and/or Customer's compliance with this Agreement.

12.2   Assignment. This Agreement will be binding upon and inure to the benefit of the parties to this Agreement and their respective successors and permitted assigns, provided that neither this Agreement nor any license hereunder may be assigned by Customer (whether by operation of law or otherwise) without Provider's prior written consent. Notwithstanding the foregoing, Provider may assign all or any part of its rights and obligations under this Agreement to (a) any entity resulting from any merger, consolidation or other reorganization of Provider, (b) any operating entity controlling Provider, or owned or controlled, directly or indirectly, by Provider, (c) any affiliate of Provider, or (d) any purchaser of all or substantially all of the Provider's assets.

12.3   Force Majeure. Neither party may be held liable for any damages or penalty for delay in the performance of its obligations hereunder (other than Customer's obligation to make payments under this Agreement) when the delay is due to the elements, acts of God or other causes beyond its reasonable control.

12.4   Arbitration. The parties agree that all disputes arising out of or relating to the formation, performance or alleged breach of this Agreement will be determined and settled by binding arbitration to take place exclusively in accordance with the commercial rules of the American Arbitration Association. Any award rendered shall be final and binding on the parties, and may he entered as a judgment by any court of competent jurisdiction. The prevailing party shall be entitled to recover its costs of arbitration (including reasonable attorneys' fees), which will he made a part of the arbitrator's award. Notwithstanding the foregoing, in the event irreparable injury is shown, either party may obtain injunctive relief exclusively in the appropriate state or federal court in [insert location of court]. Any litigation between the parties, including litigation to enforce an arbitration award, will take place exclusively in the appropriate state or federal court in [insert location of court].

12.5   Severability. A determination that any provision of this Agreement is invalid or unenforceable will not affect the validity or enforceability of any other part of this Agreement. Similarly, a determination that any provision is invalid or unenforceable in one application will not affect the validity or enforceability of the same provision in other contexts. To the extent possible, this Agreement shall be construed to give meaning to every provision.

12.6   Waiver and Modification. A party's waiver of any breach or its failure to enforce any term of this Agreement may not be deemed a waiver of any other breach or of its right to enforce the same term or others in the future. Any waiver, amendment, supplementation or other modification or supplementation of any provision of this Agreement will be effective only if in writing and signed by both parties.

12.7   Governing Law. This Agreement will be governed by and construed in accordance with the substantive laws of the United States and the State of [insert state], without regard to or application of [insert state]'s conflicts of law rules. This Agreement will not be governed by the United Nations Convention on the International Sale of Goods or the Uniform Computer Information Transactions Act, the application of which are expressly excluded.

12.8   Notices. All notices required or permitted under this Agreement must be in writing, must reference this Agreement and will be deemed given: (i) when sent by facsimile with a confirmation page generated by the sending device; (ii) 5 business days after having been sent by registered or certified mail, return receipt requested, postage prepaid; or (iii) I working day after deposit with a commercial overnight carrier, with written verification of receipt. To be effective, a confirmation copy of a notice must be sent contemporaneously via U.S. mail. All communications must be sent to the contact information set forth below or to such other contact information as may be designated by a party by giving written notice to the other party pursuant to this section 12.8:

To Provider: To Customer:

_____   _____

_____   _____

_____   _____

Attn: Legal Notices   Attn: Legal Notices

Phone: _____   Phone: _____

Fax: _____   Fax: _____

12.9   Relationship of Parties. This Agreement will not be construed as creating an agency, partnership, joint venture or any other form of legal association between the parties and each party is an independent contractor.

12.10   Attorneys' Fees. In any action or proceeding to enforce rights under this Agreement, the prevailing party will he entitled to recover its costs and attorneys' fees.

12.11   Construction. This Agreement shall be deemed the joint work product of the parties and may not be construed against either party as drafter. Captions are for convenience only and may not be construed to define, limit or affect the construction or interpretation of this Agreement.

12.12   Entire Agreement. This Agreement, including Exhibits and any order form or Statement of Work, constitutes the entire agreement between the parties with respect to the subject matter of this Agreement, and supersedes and replaces all prior or contemporaneous written or oral statements, understandings or agreements. Except where otherwise set forth in the main body of this Agreement, in the event of a conflict between an Exhibit or Statement of Work and the main body of this Agreement, the terms of the main body of this Agreement will prevail.

Provider   Customer

By _____   By _____

Name _____   Name _____

Title _____   Title _____

Date _____   Date _____

Exhibit A
Services and Support Ordered by Customer

1. *[Name of Services] Ordered by Customer*

Service	Pricing	Quantity	Total Fees

2. *Initiation Fee*

$_____, due and payable no later than 30 days after the Effective Date.

3. *Support Fee*

Initial Support Level Purchased (Check One):

   ☐ Standard Support
   ☐ Gold Support

Support Fee:

$_____ for service through dd/mm/yyyy. Thereafter, support fees will be at Provider's current rates. Provider's support fees as of the Effective Date for the Services purchased by Customer are $_____/ per calendar quarter, but are subject to change upon 30 days notice from Provider. Customer shall pay all support fees in advance, no less than 30 days prior to the first day of the calendar quarter for which the support services are purchased.

Change Fee for Downgrade in Support Level:

$_____

4. *Locations*

Customer shall be authorized to access the Services from the following locations::_____.

Exhibit B
Support Terms and Conditions

Provider offers different levels of support services, for which it charges different fees. Provider reserves the right to change the level or quality of support upon thirty (30) days written notice.

1.   Support Provided

a.   Support Hours. Provider will provide telephone consultation and advice to Customer and respond to e-mail and fax messages sent to Provider's support department: (i) for standard support, between the hours of [insert hour] a.m. and [insert hour] p.m., Pacific Time, Monday through Friday, excluding holidays; and (ii) for gold support, between the hours of [insert hour] a.m. and [insert hour] p.m., Central Time, Monday through Saturday, excluding holidays.

b.   Limitations. Customer acknowledges that Provider is under no obligation to support any hardware or software that is not part of the Service. Response times are based upon the severity level of Customer's problem, as determined by Provider and described in Section 2 of this Exhibit B.

2.   Service Guidelines

a.   Problem Classification. Provider will assign a severity level to each problem reported by Customer, based on the Problem Classification Table below. Severity 3 will be the default severity level, unless otherwise specified by Provider's support personnel. Severity levels are defined as follows:

Problem Classification Table

Problem Classification	Criteria
Severity 1 (Critical)	The Service is completely unavailable or unusable. The problem affects time-critical applications, without which Customer cannot conduct business. No known work-around is currently available.
Severity 2 (Serious)	The Service is significantly impaired. Customer cannot conduct its integral business. No known work-around is currently available.
Severity 3 (Degraded)	The Service is not functioning in accordance with its specifications. Integral business processes of Customer, however, have not been interrupted.
Severity 4 (Minimal)	The Service problems have little or no impact on Customer's daily business process.

b.   Error Reporting and Response. For all (i) e-mail and fax support and (ii) Customer problems which cannot be resolved in an initial telephone support conversation, Customer must provide Provider with a reasonably detailed description of the problem by e-mail or fax. Provider will take the following steps, in accordance with the Response Expectation Tables below:

Step 1 Provider will acknowledge Customer's problem and begin collecting additional information from Customer.

Step 2 Provider will actively address the problem during support hours (as specified above) and provide a temporary patch, correction, or workaround as soon as reasonably possible.

Step 3 Provider will provide a permanent solution in the form of a tested permanent patch or a completely new release of the applicable software.

Customer must provide Provider with contact information for employees who will be available outside of Customer's normal business hours if any support will be provided at such times.

Provider will use reasonable efforts to respond to Customer's support problems during the response times indicated in the Response Expectation Table. In most cases, Provider will respond as follows:

Response Expectation Table (Standard Support)

Severity	Step 1	Step 2	Step 3
1 (Critical)	4 support hours	Immediate and continuing effort during support hours	Within 60 calendar days
2 (Serious)	8 support hours	1 to 5 business days	Within 90 calendar days
3 (Degraded)	16 support hours	Within 10 business days	Next scheduled release of software
4 (Minimal)	24 support hours	Worked on a time available basis	If Provider determines this step to be necessary

Response Expectation Table (Gold Support)

Severity	Step 1	Step 2	Step 3
1 (Critical)	2 support hours	Immediate and continuing effort 24 hours a day	Within 30 calendar days
2 (Serious)	4 support hours	2 business days	Within 45 calendar days
3 (Degraded)	8 support hours	Within 5 business days	Next scheduled release of software
4 (Minimal)	16 support hours	Worked on a time available basis	If Provider determines this step to be necessary

c.   Escalation Process. If the Provider support representative on duty at the time Customer reports a problem cannot correct any level 1 to 3 problem or implement a plan of resolution within the time set forth in the Escalation Table below, (s)he will notify support management, which will do so. At each stage in the Escalation Table below, Customer shall make available to Provider a Customer contact at an equivalent management level, who has the authority to make decisions about alternative approaches for resolving Customer's problem. In addition, for level 1 problems, Customer shall make available to Provider a Customer contact who will be continuously available to assist Provider's support personnel with data gathering, testing, and applying fixes. Upon Provider's request, Customer shall provide access to its computing environment if Provider cannot duplicate Customer's problem in-house.

Provider will make reasonable efforts to notify its support management, according to the following schedule:

Escalation Table

Elapsed Time	Severity 1 (Critical)	Severity 2 (Serious)	Severity 3 (Degraded Operations)
Immediately	Support Group Leader		
2 hours	Support Manager		
4 hours	Support Director	Support Group Leader	
8 hours			
16 hours	VP of Development	Support Manager	
24 hours			Support Group Leader
48 hours		Support Director	
72 hours	CEO	VP of Development	Support Manager

Elapsed Time represents the number of support hours (not clock hours) that have passed since the problem was reported by Customer.

Exhibit C
Consulting Services Order Form

This Order is entered into as of the __ day of _____, 20__, between _____("Customer") and _____ ("Provider").

Whereas, Customer and Provider executed a Master Agreement having an Effective Date of _____ ("Agreement"), for the acquisition of computer products and/or related services from Provider by Customer;

Now, therefore, in consideration of the mutual covenants, agreements, and conditions contained herein, Customer and Provider agree as follows:

1.   PRODUCTS/SERVICES:

QTY.	TYPE	PRICE/ LICENSE FEE	ANNUAL MAINT/ SUPPORT CHARGE	DELIVERY /SERVICES DATE	INSTALL DATE

2.   LOCATION:

The equipment/software/services shall be delivered/installed/performed at the following location:

Ship to:	Bill to:
Attn: _____	Attn: _____

3. TECHNICAL PERSONNEL: All   technical   personnel contracted to Customer by Provider that are deemed by Customer to be unsatisfactory for any reason will be immediately removed by Provider. Provider shall notify Customer, within 24 hours, of either a removal or discontinuance, whether and when a suitable replacement will be provided, or whether Provider will be unable to provide a suitable replacement. Customer, in its sole discretion, shall have the right to accept or reject any replacement personnel offered by Provider. Customer shall provide on-site technical personnel with adequate work space, and necessary clerical services and incidental supplies at no cost to Provider.

4. ORDER TERMINATION BY CUSTOMER: Customer may terminate any Services pursuant to this Order, with or without cause, by giving 30 days prior written notice to Provider.

CUSTOMER   (Provider)

BY: _____   BY: _____

NAME: _____   NAME: _____

TITLE: _____   TITLE: _____

DATE: _____   DATE: _____

ABOUT THE AUTHORS

Kelly L. Frey, Sr.

Kelly Frey concentrates his practice in the area of corporate and information technology law. He represents the headquarters locations of several Fortune 500 companies and regularly advises clients in purchasing and selling IT products/services, as wells as intellectual property (IP) matters (including copyright, trademark, trade secrect, domain name, Internet, and commerce issues). Mr. Frey has extensive experience in negotiating and drafting confidentiality agreements, term sheets, letters of intent/memoranda of understanding, software development and licensing agreements, business process/IT outsourcing agreements, ASP/SaaS agreements, consulting agreements, and content licensing agreements (for both traditional and new media). He is the author of the legal practice guide *Frey on Intellectual Property and Technology Transactions*. As a result of having spent ha;f of his professional career as in-house counsel, Mr. Frey understands the needs of his coportate clients, and his emphasis is on assisting clients chieve their practical business goals on time and on budget.

Thomas J. Hall

Thomas Hall is of counsel in the Baker Donelson's Nashville office and is a member of the Firm's Business Law Department. He concentrates his practice in the area of corporate and information technology law. He has twenty years of experience in negotiating and drafting confidentiality and nondisclosure agreements, term sheets, letters of intent/memoranda of understanding, software development and licensing agreements, IT outsourcing agreements, ASP agreements, consulting agreements, patent licenses, joint development agreements, and supply and distribution agreements. Mr. Hall has drafted and implemented template agreements for software, hardware, telecomm, consulting, confidentiality, and e-commerce for various clients.

The authors have donated all royalties for this publication to the Susan G. Komen Breast Cancer Foundation in memory of Dove Hall, Sue Porter, William and Louise Frey, and F. B. and Henry Stewart (all victims of cancer).